Oxford International Primary Science

5

Terry Hudson

Alan Haigh

Debbie Roberts

Geraldine Shaw

Language consultants:
John McMahon
Liz McMahon

OXFORD
UNIVERSITY PRESS

Great Clarendon Street, Oxford, OX2 6DP, United Kingdom

Oxford University Press is a department of the University of Oxford. It furthers the University's objective of excellence in research, scholarship, and education by publishing worldwide. Oxford is a registered trade mark of Oxford University Press in the UK and in certain other countries

First published in 2016

e-Book Edition
9780198413493 e-Book
9780198413554 e-Book (In-App)

Acknowledgements
The publishers would like to thank the following for permissions to use their photographs:

Cover photo: Paul Souders/Corbis, P04_05: Katrina Brown/ OUP, P5: Ozerov Alexander/Shutterstock, P7: Stockbyte/OUP, P8: Philip Coblentz/OUP,P 10a: Tom Wang/Shutterstock, P10b: Shutterstock, P10c: Fabio Alcini/Shutterstock, P10d: Corbis/OUP, P10e: Shutterstock, P10f: Ilya Shapovalov/Shutterstock,P10g: Artazum and Iriana Shiyan/Shutterstock, P10h: Shutterstock, P12: Image Source/OUP, P13a: Photodisc/OUP, P13b: Shutterstock, P14: Alaska Stock Images/ National Geographic Creative, P16: Donald Iain Smith/Moment/Getty Images, P19: Dinodia Photos/Alamy, P20: Fotolia, P22_23: Kim Walker/Robert Harding World Imagery/Corbis/Image Library, P23a: Christophe Boisvieux/Corbis/Image Library, P23b: Shutterstock, P23c: Shutterstock, P24: Dreamstime.com, P25a: Gina Smith/Dreamstime.com, P25b: Shutterstock, P29a: Alex Staroseltsev/ Visual Photos, P29b: Shutterstock, P29c: Dreamstime.com, P30a: Peter Gudella/ Shutterstock, P30b: Shutterstock, P31a: Shutterstock, P31b: Sayuri Mori/ Getty Images, P32a: Shutterstock, 32b: Shutterstock, P33: Lee Snider Dreamstime. com, P34a: Shutterstock, P34b: Shutterstock, P36: Africa Studio/Shutterstock, P37: Alexander Yu. Zotov/Shutterstock, P39: Peter Mason/cultura/Corbis/ Image Library, P41: Steven Coling/Shutterstock, P42a: Martyn F. Chillmaid/Science Photo Library, P42b: Feig/Feig/the food passionates/Corbis/Image Library, P43: Galen Rowell/Corbis/Image Library, P46: Shutterstock, P48: Shutterstock, P50a: Vaughan Fleming/Science Photo Library, 50b: Viktor Fischer/OUP, P50c: Shutterstock, P51a: Javier Trueba/Science Photo Library, P51b: David Reilly/Shutterstock, P53a: Dreamstime.com, P53b: Justin Sullivan/Getty Images News/Getty Images, P54a: Shutterstock, P54b: Shutterstock, P54c: Shutterstock, P54d: Shutterstock, P54e: Jan Kaliciak/Shutterstock, P55a: M. Unal Ozmen/ Shutterstock, P55b: Marie C Fields/Shutterstock, P56a: Lemonakis Antonis/Shutterstock, P56b: Shutterstock, P57: Shutterstock, P58: Shutterstock, P60_61: Shutterstock, P61a: Maks Narodenko/Shutterstock, P61b: Fotolia, P61c: Shutterstock, P61d: Arvind Balaraman/Shutterstock, P62a: Fotolia, P62b: Maximilian Stock td/ Shutterstock, P62c: Shutterstock, P62d: Richard Griffin/Shutterstock, P62e: Michael Flippo/Dreamstime.com, P63a: Arvind Balaraman/Shutterstock, P63b: F otolia, P63c: Carolyn Franks/ Dreamstime.com, P63d: Shutterstock, P63e: Maks Narodenko/Shutterstock, P64: Richard Griffin/Shutterstock, P67a: S.J. Krasemann/ Photolibrary/Getty Images, P67b: Brian A Jackson/Shutterstock, P67c: Itrendo Nature/Shutterstock, P67d: Dreamstime.com, P68: Dreamstime.com, P69a: Stephan Morris/Shutterstock, P69b: Korovina Daria/Shutterstock, P69c: Brian A Jackson/Shutterstock, P69d: Shutterstock, P69e: Luke Wein/Shutterstock, P69f: Susan Leggett/Shutterstock, P70: Christian Musat/Shutterstock, P71a: Linda Burgess/Photolibrary/Shutterstock, P71b: Robert & Jean Pollock/Visuals Unlimited/Corbis/Image Library, P71c: Aravind/Flickr Open/Getty Images, P72a: Masterfile, P72b: Shutterstock, P73: Shutterstock, P84_85: Jarno Gonzalez Zarraonandia/Shutterstock, P85: Shutterstock, P86a: iStock.com, P86b: Shutterstock, P86c: Lane V. Erickson/Shutterstock, P86d: Shutterstock, P86e: Eniko Balogh/Shutterstock, P86f: Shutterstock, P87a: Jeremy Reddington/Shutterstock, P87b: Shutterstock, P87c: Christopher Wood/Shutterstock, P88: Shutterstock, P90: Dirk Ott/Shutterstock, P91: Shutterstock, P92: Roman Slavik/Shutterstock, P93: iStock.com, P94: Shutterstock, P98a: Petr Salinger/Shutterstock, P98b: Uros Ravbar/Dreamstime.com, P100_101: Shutterstock, P101a: NASA/ESA/Hubble/HPOW, P101b: NASA/Science Photo Library, P104: Pekka Parviainen/Science Photo Library, P112: Shutterstock, P114: Panos Karas/Shutterstock, P115: Aleksey Tugolukov/123RF, P116: Matthias Haas/Fotolia, P117: Reto Stöckli, Nazmi El Saleous, and Marit Jentoft-Nilsen/NASA/ GSFC, P120_121: Thomas Winz/Lonely Planet Images/Getty Images, P121: Adam J/Shutterstock, P122a: Shutterstock, P122b: Franck Boston/Shutterstock, P122c: Shutterstock, P122d David Butow/Corbis/Image Library, P122e: Cameron Davidson/Corbis/Image Library, P122d: Shutterstock, P124: Shutterstock, P125: Photodisc/OUP, P126a: Shutterstock, P126b: Yin Dongxun/Xinhua Press/Corbis/Image Library, P127: Natalia Sheinkin/ Shutterstock, P131: Michel Gouverneur/Reporters/Redux, P132: Ingram/OUP, P134: Heintje Joseph T. Lee/Shutterstock, P135a: Ocean/OUP, P135b: Alexander Kalina/Shutterstock, P135c: Shutterstock, P136a: Popperfoto/Getty Images, P136b: Robert Simmon/NASA, P142a: Itrendo Nature/Shutterstock.

Although we have made every effort to trace and contact all copyright holders before publication this has not been possible in all cases. If notified, the publisher will rectify any errors or omissions at the earliest opportunity.

Contents

How to be a Scientist

Scientists wonder how things work. They try to find out about the world around them. They do this by using scientific enquiry.

The diagram shows the important ideas about scientific enquiry.

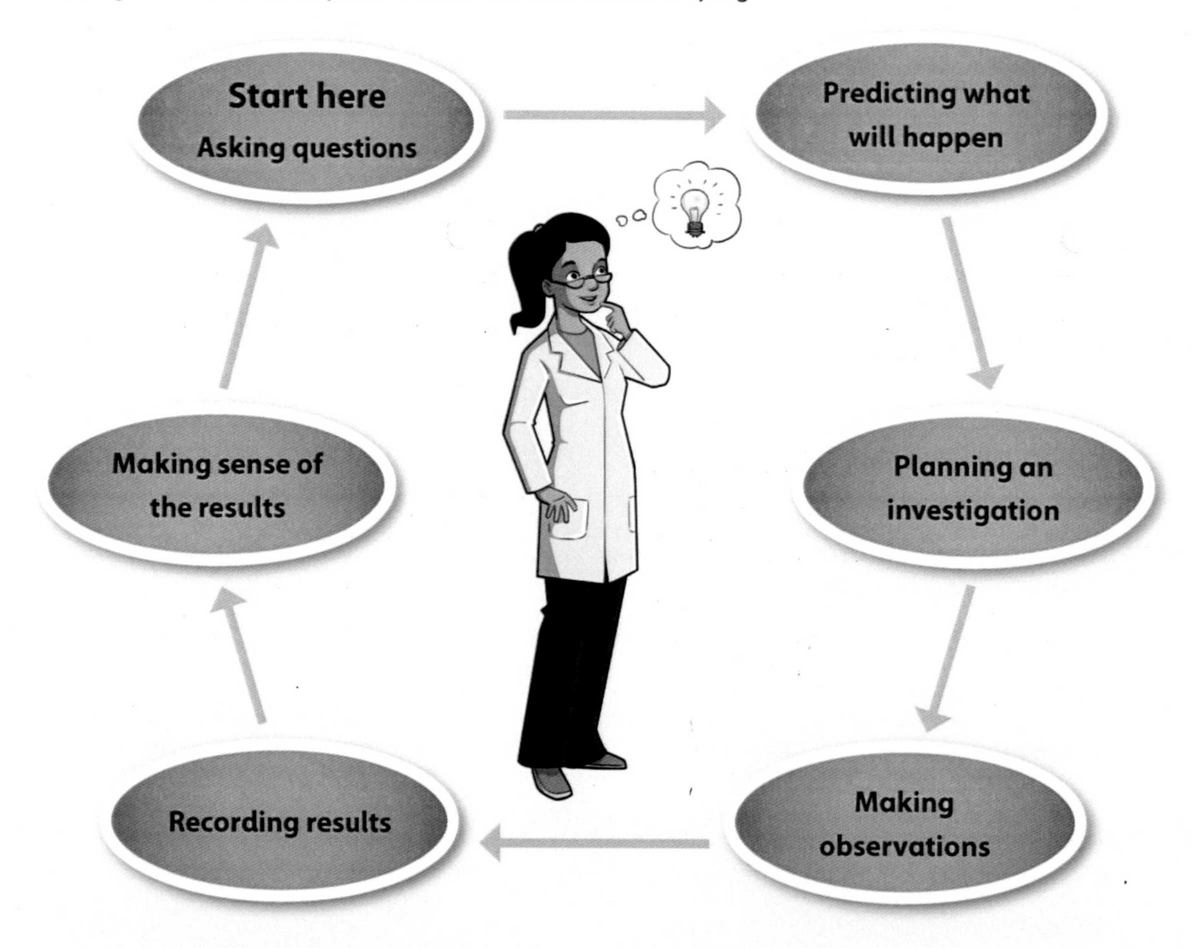

An example investigation: Which materials make the best shadows?

Asking questions

How can you ask questions?

Start your questions with words like 'which', 'what', 'do' and 'does'.

- Which materials make the darkest shadows?
- Are there some materials which do not make shadows?
- Does the darkest material make the darkest shadow?

Predicting what will happen

A prediction is when you say what you think will happen in your investigation. To stop a prediction being a guess try to give a reason. For example:

Question

Do transparent materials make very good shadows?

Prediction

No.

Reason

Shadows are made when light is blocked. Transparent materials let light pass through. A shadow will not be made.

Planning an investigation

When you plan an investigation think about how you are going to set it up and how you are going to do it. What equipment will you need?

You will also need to make it a fair test.

What will you keep the same?

- The amount of material you are testing.
- The source and brightness of light you are using.
- The distance between the light and the material you are testing.

What will you change?

- The type of material.

Making observations

You will need to look carefully at the shadows being made.

To make your investigation accurate you will measure how far away from the light source you hold the materials. Use a ruler.

You will look at the shadow and record how dark it is for each material.

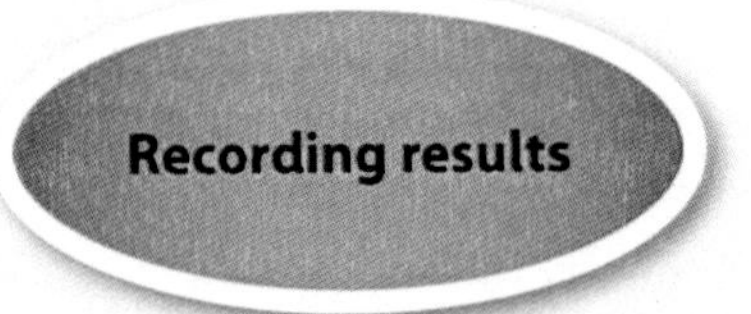

There are many ways to record results. A good way is to complete a table. You can also draw or photograph the shadows. A table keeps results neat and tidy. It can help you to see patterns. How else you could show your results? Would a chart or graph be helpful?

Making sense of the results

At the end of your investigation you must look at your results carefully. Check to see which materials made the darkest shadows.

Was your prediction correct? Can you think of ways of making your investigation better? What next?

Scientific enquiry always leads on to other questions and more investigations. For example: How can we change the size of shadows?

1 The Way We See Things

In this module you will:

- build on your understanding of light from earlier work
- explore how light can be reflected from surfaces
- find out about how mirrors work and why they are very useful to us.

Amazing fact

The Sun is approximately 150 million kilometres away, but it only takes the light 8 minutes and 20 seconds to reach our eyes.

The Sun is our main source of natural light. The Sun gives off so much light it can heat and light up whole planets.

Light is the fastest thing that we know about. In one second a beam of light can travel around the world approximately seven times.

Can you race a beam of light across a room?

Sunlight is actually white. It is hard to believe, but it contains all the colours of the rainbow mixed together. Raindrops can split the light into the different colours.

Have you ever wondered what a rainbow is?

How do our eyes see things?

Know that we see light because light from the light source enters our eyes.

The Big Idea

Without light our eyes cannot see anything.

Can you remember how light travels? Write down two facts.

it reflects of of things

Think back to how light travels from a source to our eyes.

List three **light sources** that you have seen today.

sun, lamps, lighthing

Some of the light sources are natural. The Sun is a good example. Other sources of light come from **objects** that people have made. These are called artificial sources of light.

We are going to use an artificial source of light to investigate some objects. The light source is a torch.

Investigation: Light shining on objects

Use your torch to investigate objects around the room. Shine the light from the torch on to the different objects. Write down which objects bounce the light back.

When light is bounced back from an object, we say it has been **reflected**. An object that does this well is called a good reflector; an object which does this badly is called a poor reflector.

Look at the results from your torch investigation. Group the objects into good reflectors and poor reflectors. In your Investigation Notebook, copy and complete the table below.

Good reflectors	Poor reflectors
miorors	books
moon	rulars

Choose the object that you think is the best reflector out of the ones you tested.

What kind of material is it? Describe the properties of the material. Would you describe the material as shiny or dull?

If something is shiny, it reflects light because it is very clean or polished; if it is dull, it is not bright or clean and does not reflect light.

Remember

There has to be a source of light for us to see things.

The Moon is not a source of light. It is like a giant **mirror** in the sky. The Moon does not make its own light. It reflects it from the Sun.

How did the Moon landings prove to us that the Moon is not a light source?

because its reflecting

What do you think the far side of the Moon looks like?

gray and full of cratars

If we look at a light source, the light is travelling directly into our eyes.

⚠ Do not do this with a bright light source such as the Sun. It can damage your eyes.

When we look at objects that are not light sources, we see them in a different way. The light is not made by the object. It is reflected from the object.

1 Circle one of the following that is not a source of light.

Sun **candle** 

torch **lamp**

2 Write the name of a material that reflects a lot of light.

miorors

3 When light bounces off an object it is

r. reflecting

Now turn to page 20 to review and reflect on what you have learned.

How are shadows made?

Be able to identify shadows.

The Big Idea

Shadows are made when light is blocked.

In your Investigation Notebook, construct a mind map like the one shown below. Include everything you can remember about shadows. You can add as many boxes as you need.

Shadows

Think back to your previous learning about shadows.

How are shadows formed?

They are forment when light gets blocked.

Remember
Light travels in a straight line. It travels at 1 billion metres every second.

If an object is in the path of a **beam** of light, the light beam will stop or bounce back. The beam of light cannot go around the object. This is how shadows are formed.

Some objects let light pass through them. We say these objects are **transparent**. Other objects let some light through and these are **translucent**. An object that completely blocks the light is **opaque**. Opaque objects form the best shadows.

Investigation: Objects and shadows

Look around the room. Find five small objects.

Think about:

- what equipment you will need
- what you will use as a source of light
- how you will know which is a good shadow and which is not.

How would you investigate the five objects to find out if they make shadows? Plan an investigation to test your objects.

Find out what happens as you move the light source closer to the object. Think about how the shadow changes.

Write down what you find out about the size of the shadow as you move the light source nearer to the object.

the shadow gets bigger when you get closer.

How reliable are your results?

What type of shadow does a translucent material make?

What type of shadow does a transparent object make?

Think about...

1 What would a world without reflection be like?

2 What problems would we face if light did not reflect off some objects?

Decide whether the following statements are true or false.

		True/False
1	Light travels in straight lines.	false
2	When light hits a transparent material it passes through.	True
3	When light hits an opaque material it can pass through.	false
4	Shadows are formed when light passes through a transparent material.	True

Did you know?

When scientists carry out investigations they collect results. To be absolutely sure they have collected reliable results they repeat the investigation many times.

Now turn to page 20 to review and reflect on what you have learned.

The journey of light

Know that beams/rays of light can be reflected by surfaces including mirrors into our eyes.

Know that reflected light enters our eyes and we see the object.

The Big Idea

Mirrors have many important uses. Some are even lifesaving.

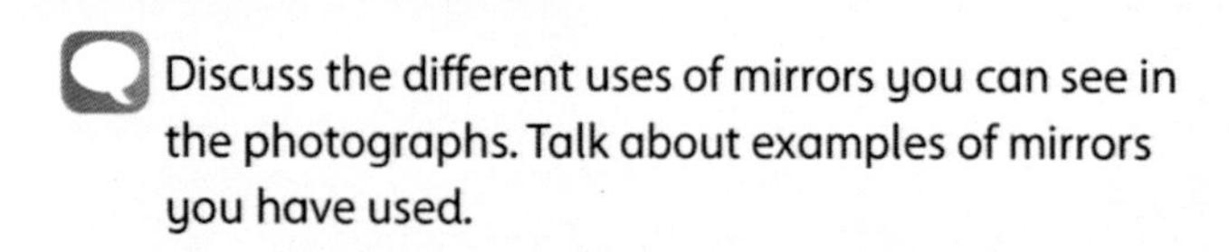

Discuss the different uses of mirrors you can see in the photographs. Talk about examples of mirrors you have used.

Remember

Light is reflected from some **surfaces** better than others.

Amazing fact

The image you see of yourself in a mirror is back to front (the wrong way around).

Can you remember two types of material which reflect light more than others?

mirrors, windows

Investigation: The mirrored image

How can you prove that the mirrored image is back to front?

Place an object in front of a mirror. Point to the left side of the object and look in the mirror. What do you see?

It's on the right side.

Point to the right side of the object and look in the mirror. What do you see?

It's on the left side

Follow the path of Star A with a pencil. Draw between the two outer lines and do not touch the sides. Now collect a small mirror. Look at Star B in the mirror and try to draw around it between the two lines. Do not look at Star B, only look at its mirrored image.

Was it easier to draw around Star A or Star B? Why?

Star A was easier than Star B because, I used a mirror.

Is the amazing fact, left, true or false?

true

The journey of light

Know that beams/rays of light can be reflected by surfaces including mirrors into our eyes.

Know that reflected light enters our eyes and we see the object.

The Big Idea

Mirrors help us to see behind us, round objects and over objects

Mirrors are usually made out of ordinary glass or coated plastic. The back of the mirror is covered in a shiny metal foil that is designed to reflect light very well.

Mirrors are really good reflectors. Many other shiny objects reflect light well. Dark and black objects do not reflect light very well. This is because the light beam cannot bounce off the surface as well. It is absorbed.

- Why are mirrors used on the inside and the outside of buildings?
- Which are the reflective surfaces of the building?

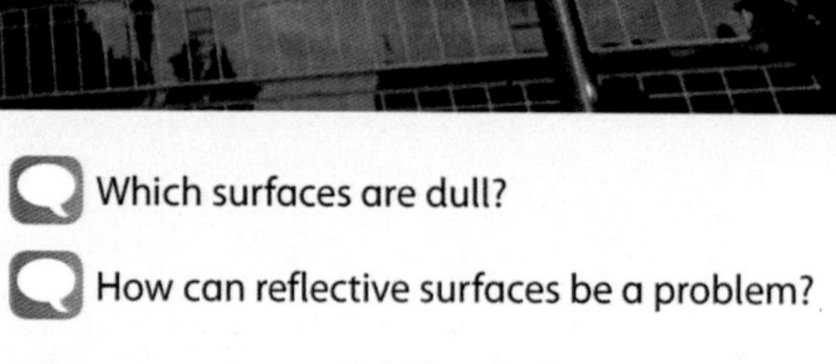

- Which surfaces are dull?
- How can reflective surfaces be a problem?

Mirrors are very useful. They help us to see things that we normally cannot. People who drive vehicles use mirrors to see behind them. You may be able to arrange some mirrors to explore how this works.

- Is it possible to use mirrors to see the back of your head?

Investigation: Periscope

Your teacher will show you how to make a periscope. This is a device for seeing around corners, or over things. When you use a periscope you can see what is happening on the other side of a wall.

More uses of mirrors

Mirrors have many other uses. Dentists and doctors use them to see into difficult places. For example, a dentist has to be able to see behind your teeth.

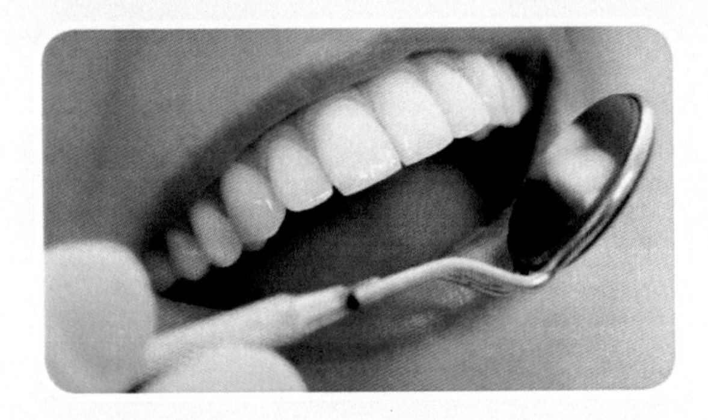

Mirrors are also used to make the light in a torch brighter. If you look closely at a torch, the surface behind the bulb is usually like a mirror to make the light more intense.

How can you use a piece of aluminium foil to make a bulb brighter?

In your Investigation Notebook, draw a diagram to show how you can set up the bulb and foil to get the brightest light. You may want to think about where you place the foil. The shape may also be important.

The journey of light

Know that beams/rays of light can be reflected by surfaces including mirrors.

Know that reflected light enters our eyes and we see the object.

The Big Idea

We can follow the journey light takes using ray diagrams.

The person in the photograph is seeing the Moon but not looking directly at it. How is light from the Moon travelling to the person's eyes?

it's reflecting

The Moon is not a light source. Why does it look bright?

the sun ray reflects it

Scientists show the direction of light by drawing **ray** diagrams. The straight lines show the beam or ray of light and the arrows show the direction the light is travelling in.

If you shine a torch on to a dark surface, you can see the beam of light. The beam leaves the torch and travels through the air until it hits the surface.

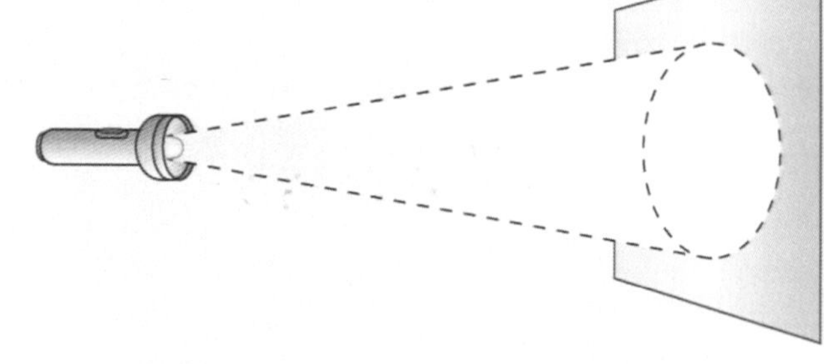

Scientists use ray boxes to investigate light.

Remember

Light travels in a straight line. You can prove this using your ray box.

 Investigation: Making a ray box

You can make your own ray box using a torch and a small box such as a shoe box.

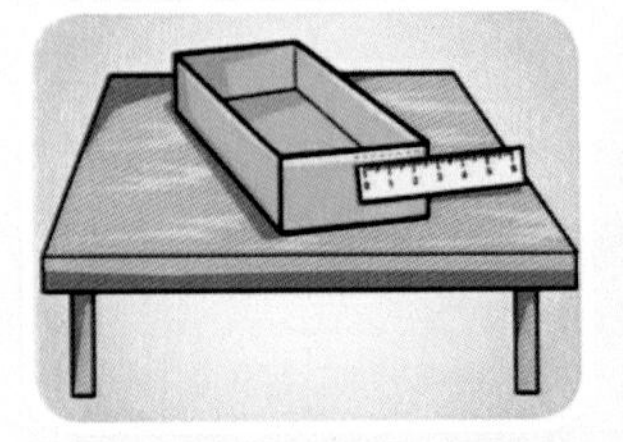

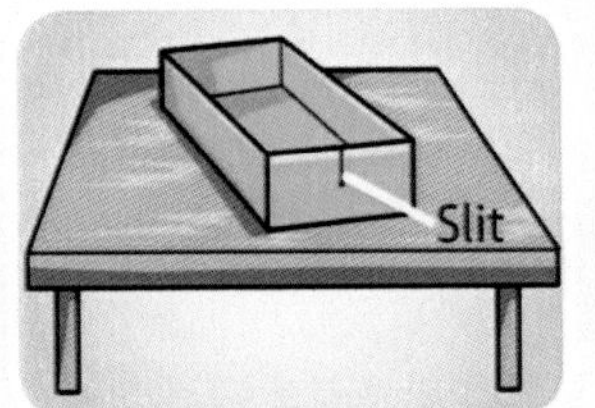

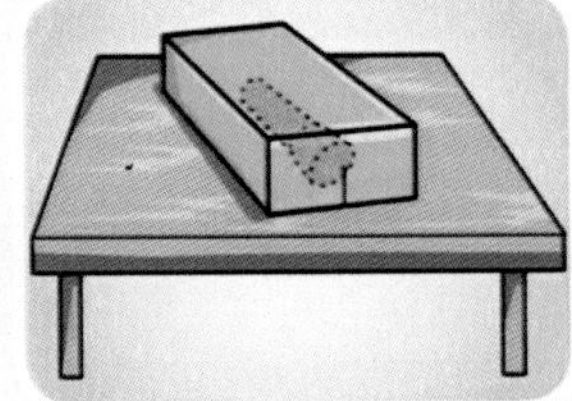

1 Take your box and use a ruler to find the centre of one of the short sides. Cut a slit from the open side of the box towards the centre.

2 Place your lit torch on the table. Place your ray box bottom side up over the torch so that the slit touches the table surface.

3 Darken the room. Move your ray box until you get a long, thin beam of light shining out through the slit.

4 Shine the light across a piece of paper and draw along the path of the beam. Using a ruler will help you to do this neatly. You have drawn a ray diagram.

5 Now place a mirror in front of the ray of light.

 What happens to the ray of light when it reaches the mirror?

The mirror gets brighter

6 Draw a line along the reflected beam of light to show the ray diagram you make when using a mirror.

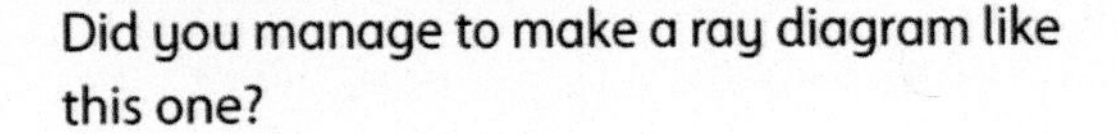

Did you manage to make a ray diagram like this one?

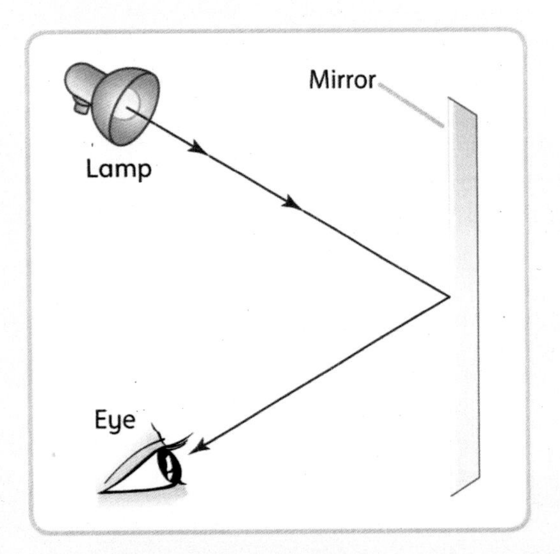

 Explain how light is reflected into your eyes when you look in a mirror.

Light bounces of the mirror into your eyes. When

Now turn to page 21 to review and reflect on what you have learned.

Light changing direction

Explore why a beam of light changes direction when it is reflected from a surface.

The Big Idea

Light travels in straight lines but can change direction.

Think back to the ray box you made. When you used the flat mirror, what happened to the ray of light?

Flat mirrors reflect a really good image called a true likeness of objects. This is because the smooth surface does not scatter the light. If we stand in front of a mirror, we see ourselves. The light reflected from us will travel to the mirror and be bounced back or reflected into our eyes.

What happens if you look into a mirror from the side? Do you see yourself? What do you see? Where do you have to stand to be able to see yourself?

Investigation: Mirrors

Use your ray box to investigate mirrors in more detail.

1 Set up the investigation so that you can draw a ray diagram of your ray of light reflecting off the mirror. This time, draw a line to show where the mirror is.

2 Draw three different ray diagrams by moving your ray box. Make sure you know which ray is which. Use a different colour for each one.

3 Take one ray diagram and draw a line at 90 degrees (a right angle) to the mirror. This is called the normal line.

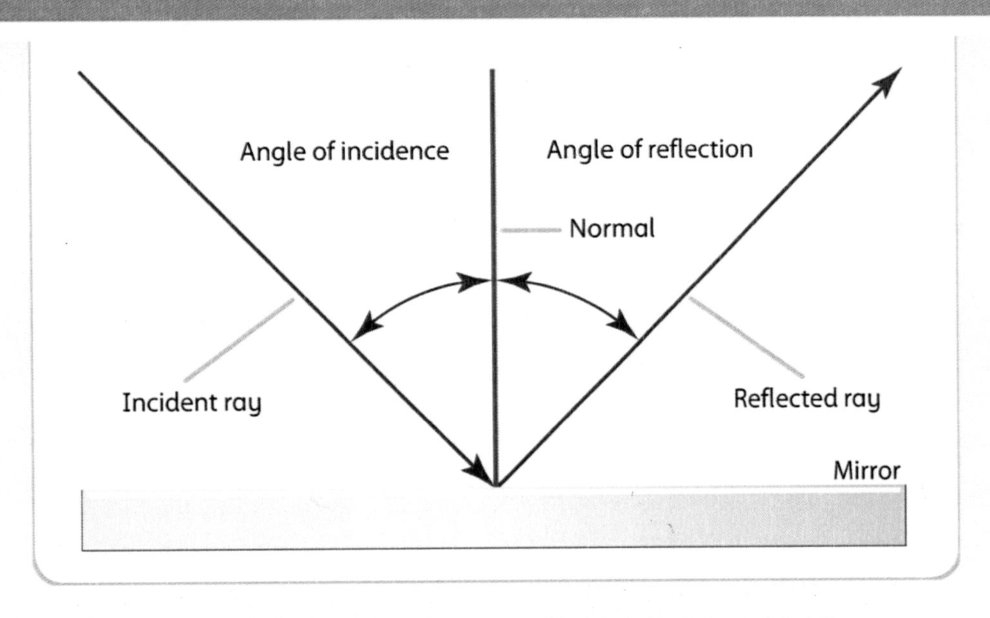

The angle at which the ray of light reaches the normal line is called the angle of incidence. The angle at which the ray of light is reflected from the normal line is called the angle of reflection. This means that the light reaching the mirror is the incident ray. The light leaving the mirror is the reflected ray.

Label the angle of incidence, angle of reflection, incident ray and reflected ray on one of your ray diagrams.

Measure the angle between the angle of incidence and the normal line. What is it?

it is 8 cm

Measure the angle between the angle of reflection and the normal line. What is it?

it is 5 cm

Measure the angle of incidence and the angle of reflection for your other two ray diagrams. What do your results show?

8 and 5 cm

1 The ray from the torch to the mirror is called the …

It is called a ray

2 The ray from the mirror to the wall is called the …

reflection

3 When light hits a mirror, what is the angle called?

Inciedent ray

4 When light bounces off a mirror, what is the angle called?

reflected ray

Now turn to page 21 to review and reflect on what you have learned.

Magnificent colours

Know that we see colour because not all of the light is reflected by some objects.

The Big Idea

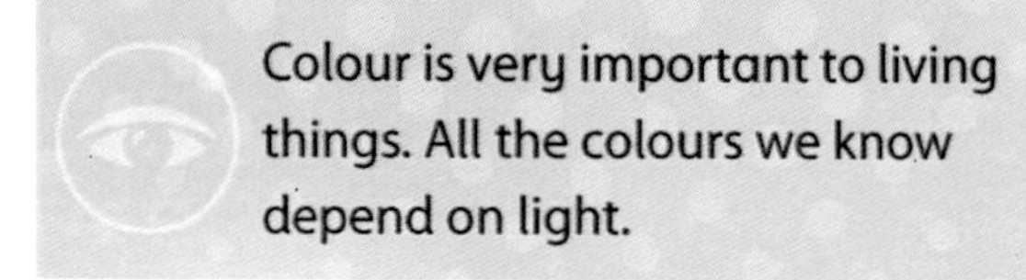

Colour is very important to living things. All the colours we know depend on light.

- Discuss all the colours you have heard of.
- What is your favourite colour? Why?
- Why do people like to have flowers around?
- Why are some places painted in bright colours?
- What happens to colours as a room becomes darker?

Remember
Some objects reflect more light than others.

Investigation: Coloured discs

1 Take a circular piece of card and divide it into eight sections. Colour each section a different colour.

2 Push a stick or pencil through the centre of the circle.

Put the card disc on the desk before pushing the stick through. Do not hold the disc in your hand.

3 Spin the stick.

What do you see?

When white light hits a surface, some of the colours are absorbed by the object and other colours are reflected. For example, a blue T-shirt absorbs most of the different colours and reflects only the blue colour.

Explain why a leaf appears green.

It apears green becouse lite refects off of it

Why do some objects appear to be white?

They apear to be white because white makes colors

Why do some objects appear to be black?

Because black is a shade

When you see a rainbow, it is because the rain droplets in the light are separating the white light into all the colours that make it.

 Never place filters over bulbs or lamps as there is a danger of fire.

Earlier you talked about your favourite colour. Explain how you can see this colour.

It is the white that reflects off of the rain droplets.

We can use different coloured filters and your ray box to mix colours of light.

 Investigation: Shadow puppets and colour

In your group, make a shadow puppet theatre. Now that you understand colour, you can use coloured filters to make the shadow puppets more interesting. Tell a story about light and colour. Don't forget to include your favourite colour.

Amazing fact

White light is made up of all the colours of the rainbow – all the colours you discussed above and more. You can prove this by spinning a colour wheel.

1 What colours make white light?

all light colors make white light

2 Explain why a red T-shirt appears red.

because light reflects off of it

3 Why does a blue top look blue in daylight but look grey when it is getting dark?

It looks like that because black absorbs color.

Now turn to page 21 to review and reflect on what you have learned.

What we have learned about the way we see things

How do our eyes see things?

(pages 6–7)

You hear students discussing how we see things. One student says that our eyes emit light which shines on an object and that is how we see things.

Explain why this is not true.

I can list some examples of sources of light.

I can explain how light is reflected from objects.

I understand that we see when reflected light enters our eyes.

How are shadows made?

(pages 8–9)

Explain how this person is making the image on the wall.

I can explain how a shadow is made.

I know what opaque, translucent and transparent materials are.

The journey of light (pages 10–15)

A boy is ill in bed. The bed is placed in front of the window so he has something to look at. Across the road there is a new office block. It is made using lots of glass. The boy can see the reflection of his house on the surface of the glass.

How can the boy see his own house?

I can make predictions about how a ray of light travels. 

I can draw ray diagrams.

Light changing direction (pages 16–17)

If you drop a bouncy ball, it will bounce back in a straight line into your hand. If you throw the ball at an angle to the ground, it will bounce off at the same angle and travel in a different direction. This is what happens when light is reflected off a mirror.

What is the name for the angle of the light as it reaches the mirror?

I know why a beam of light changes direction when it is reflected. 

I know that the angle of incidence equals the angle of reflection. 

I can draw ray diagrams showing the angle of incidence and angle of reflection. 

Magnificent colours (pages 18–19)

When you see a red car, what has happened to all the other colours of the rainbow?

I know that white light is made up of different colours. 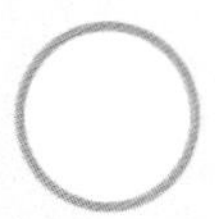

I can explain why we see certain colours. 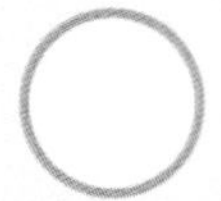

I can use different filters to change the colour of light. 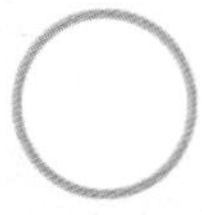

2 Evaporation and Condensation

In this module you will:

- build on what you already know about solids, liquids and gases
- find out about everyday uses of condensation and evaporation
- explain how the Earth recycles water
- investigate substances dissolving in water.

Look at the picture. Where can you see steam?

Where can you see snow and ice?

Where can you see liquid water?

Water can be a solid, a liquid or a gas.

Find a solid, a liquid and a gas in the picture.

Where else have you seen examples of these?

Is it possible to change solid ice into liquid water?

Word Cloud

boiling point
crystal
filter funnel
draw conclusions
insoluble
condensation
filter paper
evaporation
interpret data
melting point
soluble
sieve
state of matter
water cycle
water vapour
solution

- Here is a salt plain. What is being made and collected here?
- Why is the salt plain near the sea?
- Why do salt plains work best in hot countries?

This factory is making pure water from sea water. This water can then be used as drinking water for humans and animals. It can also be used to water plants.

- Why do we need factories like this to make pure water?

We can measure temperature using a thermometer.

- Why is it so important to be able to measure temperature?

Where does the water go?

Know that when a liquid turns to a gas we call it evaporation.

The Big Idea

When water warms up it seems to disappear, but it doesn't.

Someone has left a pan of water boiling on a cooker. They have forgotten all about it.

What has happened to the water?

Why is it dangerous to turn your back on something that you are heating?

its dangerous because it will make fire

When liquid water is heated, it changes into a gas called **water vapour**. This vapour spreads out into the air. If you heat water to 100°C, it will boil. This makes water vapour that is very hot. We call this steam. You must be very careful with steam because it is very hot and it can burn (scald) you.

Look at the picture of the kettle. What can you see?

steam

As the kettle boils, the water level will get lower.

Where does the water go?

it goes in the air which is called evaperation

In the summer, puddles and even rivers dry out because the heat from the Sun warms the water and it **evaporates**.

> The water doesn't disappear – it evaporates into a gas.

What do we call the gas that is made when water evaporates?

steam,

Where does this gas go?

into the atmosphier

Investigation: Evaporation and drying

Drying washing on a line keeps the clothes spread out and not folded up. This helps the water to evaporate and dry the clothes. It also helps if the weather is dry, warm and windy.

Why do spread-out clothes dry faster than folded-up clothes?

Why do clothes dry faster in hot weather than in cold weather?

Why does windy weather dry clothes faster than still, calm weather?

It is a good idea to check ideas yourself. Don't always believe what you read. This is a chance to test some ideas about evaporation and drying.

Investigate one of the questions above.

Think about:

- the different materials you will use
- how you will wet the materials at the start
- the different ways to try to dry your materials
- which things you will keep the same so it is a fair test.

Discuss some examples of liquids that you can smell when they evaporate.

Where does the water go?

Know that when a liquid turns to a gas we call it evaporation.

The Big Idea

Particles move differently in solids, liquids and gases.

List three examples of a solid, three examples of a liquid and three examples of a gas.

ice, Books, tablet, water, juce, pop, clouds, air, sun

How do you know which substances are solids, liquids and gases? Are there any rules?

The way a substance looks, feels or behaves is called a property. You know that solids, liquids and gases do not look and feel the same. They also behave differently.

Do you remember that you can draw models of how particles behave?

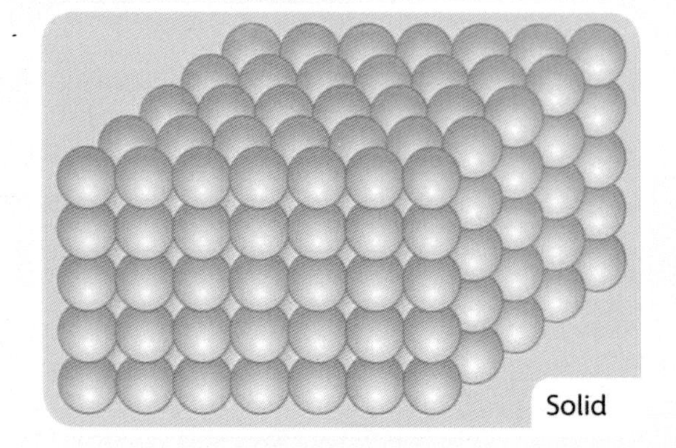

Solids are made up of particles that are tightly packed together. These particles hold on to each other. The particles vibrate but they do not move. Solids keep their shape and are often hard and durable (hard wearing).

Which properties of solids make them useful for making buildings and tools?

most of them stay a long time as that one shape

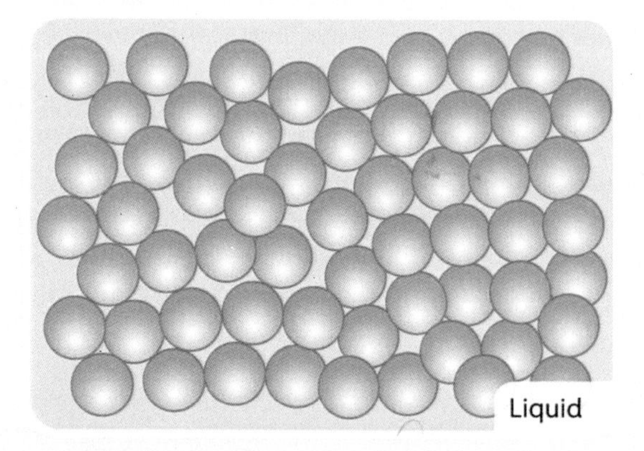

The particles in a liquid are not so tightly packed as in a solid. The particles can move more easily. This is why liquids can flow. Liquids also take on the shape of the container they are in.

Use the liquid particle model to explain why liquids take the shape of their container.

They take that shape because its partacls move aroud a lot.

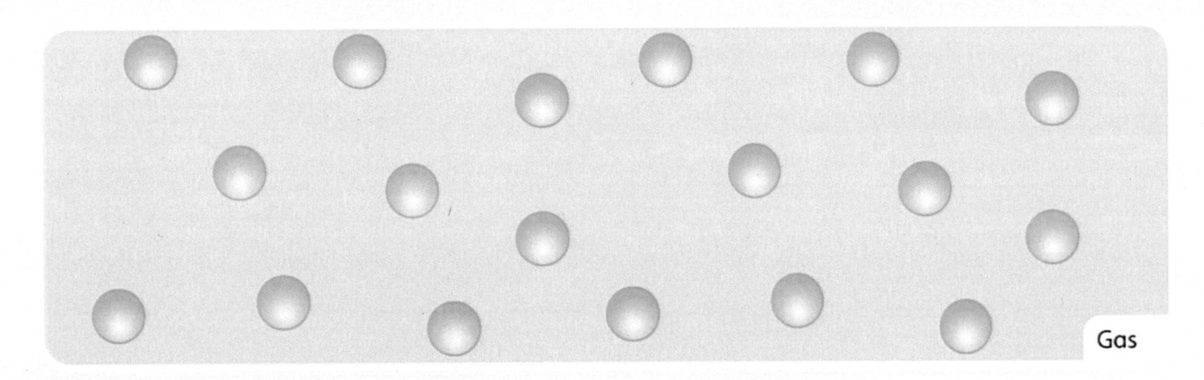

Gases are made up of particles that are widely spaced out and free to move. This means that gases can spread out easily to fill any space they are in. Unlike solids and liquids, gases have no fixed shape or volume.

Use the gas particle model to explain why gases are so easy to squash.

Is because gas is air your hand goes threw it.

Investigation: Identifying substances

Property	Solid	Liquid	Gas
Volume	Fixed	Fixed	Fills its container
Shape	Fixed	Takes the shape of the lower part of a container	Spreads out to take the shape of the whole container
Density	High	Medium	Low
Ease of compression (squashing)	Very low	Low	High
Ease of flow	No flow	Easy	Easy

Investigate different substances to identify whether they are solids, liquids or gases. Use the summary table above to help you.

You will need to plan how you will test each property. Think about how you will make your tests reliable.

Where does the water go?

Know that when a liquid turns to a gas we call it evaporation.

The Big Idea

Liquids evaporate.

Materials can change into solids, liquids or gases. Scientists call this changing the **state of matter**.

When a liquid gets cold it changes to a solid. This is called freezing. When a liquid gets hot enough the liquid can turn into a gas. This is called evaporation. When a gas cools again it turns back into a liquid. This is called **condensation**.

 Draw pictures in the boxes below to show what happens to particles as a substance changes from a solid to a liquid. This is a particle diagram.

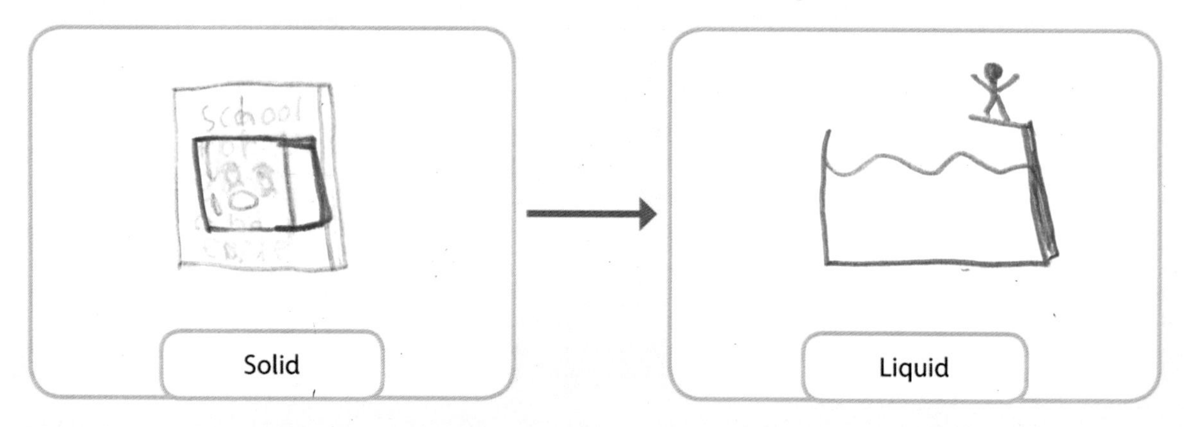

Draw pictures in the boxes below to show what happens to particles as a substance changes from a liquid to a gas.

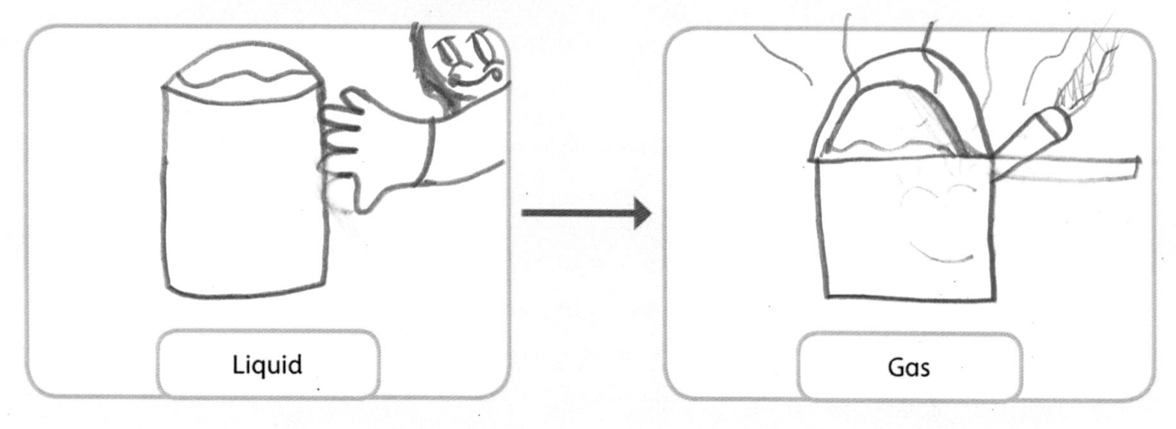

Complete the sentences using the words in the word bank below.

The change from liquid to gas is called

evaperation

The change from gas to liquid is called

condensation

The change from solid to liquid is called

____

The change from liquid to solid is called

____

Word Bank

freezing **evaporation** **melting**

condensation

Investigation: Modelling particles

You can model particles in your class. Imagine you and your classmates are particles.

1 Arrange yourself into straight rows behind each other. You must stand close together, side by side. You can move just a little.

Which state are you modelling?

What happens if the particles in this model are heated?

2 Model the movement of particles after heat has been applied. Predict how the particles behave.

Can you move now?

If your teacher continues to heat the particles, what happens?

Which new state is being made?

Have you got more or less energy?

3 Imagine your teacher takes away the heat and you get colder and colder until you freeze. Model this movement of particles.

Answer Yes or No to the following questions:

		Yes/No
1	Can you pour a large piece of solid material?	________
2	Can a gas spread out to fill a container?	________
3	Does a solid make the shape of the container it is in?	________
4	Does a liquid make the shape of the container it is in?	________

Where does the water go?

Know that when a liquid turns to a gas we call it evaporation.

The Big Idea

Steam has many uses.

 Explain where steam comes from.

 Why do we have to heat water to make steam?

As more and more steam is made, the particles squash together and hit the sides of the container. Scientists call this pressure. This pressure can be used to work machines such as steam engines.

 Where does electricity come from?

Electricity can be made in many ways. One of the most common ways of making electricity is to use steam. Water is heated using coal and steam is produced. This is just like with a steam engine, but this time the steam is used to move a turbine. A turbine is like a windmill.

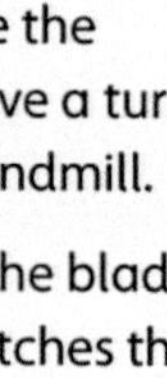

The steam catches the blades just like the wind catches the blades in a windmill. This makes electricity.

Electricity can also be made by using oil instead of coal to heat water to make steam. Ninety per cent of electricity worldwide is generated from steam today. Gases like this can cause global warming: scientists think that the Earth is warming up because we burn fuel.

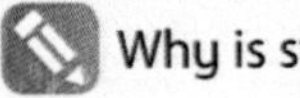 Why is steam so important to us?

 What problems are caused by burning coal and oil in power stations?

Steam can also cook food. Steam is hotter than water. This means it cooks food much quicker.

Think about...

Water is not the only liquid that evaporates. When you spray perfume on to your skin, it evaporates and that is how you can smell it.

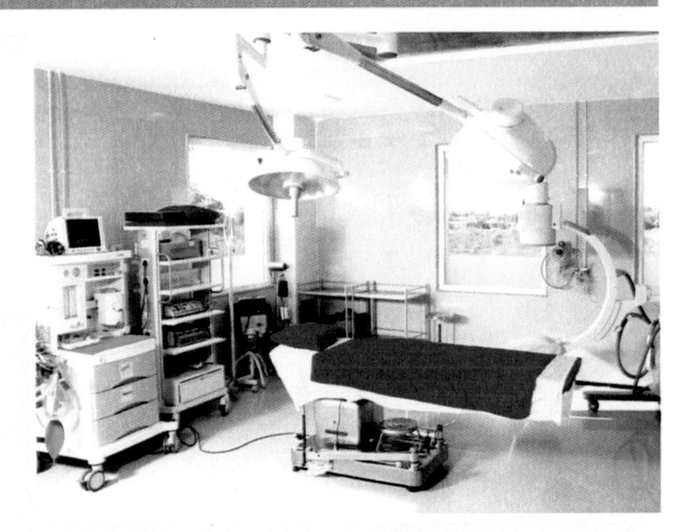

Hospitals use steam to sterilise equipment. Steam is so hot that it kills things that we cannot see, such as germs or bacteria. If you have an operation, the room and the equipment have to be very clean or they can make you ill.

In 2012, researchers began work to use solar power to heat water. The steam generated from heating the water was used to purify water. This is very important as many diseases are spread through dirty water.

 Why is using solar power to make steam better for the environment than using coal or oil?

 1 What is the scientific term for steam?

2 Circle the correct answer:

What state of matter is water vapour?

solid **liquid** **gas**

Water vapour condenses into a:

solid **liquid** **gas**

Now turn to page 58 to review and reflect on what you have learned.

Getting the water back

Know that condensation is when a gas turns back into a liquid and is the reverse of evaporation.

The Big Idea

Cooling water vapour gives us liquid water.

Remember

Water vapour is a gas. It is made when water evaporates.

What is steam?

Write four examples of where you have seen steam.

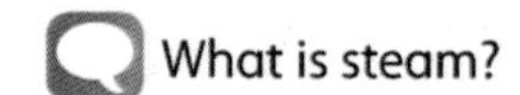

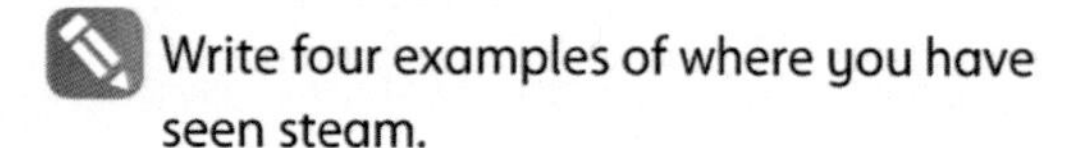

Why must you be careful not to touch steam?

You may have seen steam turning back into water on a cold surface. A gas cooling and forming a liquid is called condensation. This is the opposite of evaporation.

Look at the pictures above. Draw arrows pointing to examples of condensation.

Investigation: Condensation

Carefully hold a cold mirror in front of your mouth. Breathe out gently.

What do you see on the mirror?

See if you can make condensation on a warm mirror. In your Investigation Notebook, record your observations using a warm and a cold mirror.

 What do your results tell you about condensation?

List three places where you have seen condensation.

Look back to the picture of the pan on page 24. Now that you understand evaporation and condensation, what could you do to help keep the water in the pan?

Investigation: Slowing down evaporation

Sometimes we need to slow down the process of evaporation. For example, we can cover plants with large plastic bags to stop them from drying out.

Plan an investigation to compare seedlings growing in the open with some growing under plastic.

Think about how you will make it a fair test. Will you need to repeat measurements?

How will you measure which seeds grow the best?

How much water will you give them?

How will you find out which seeds dry out the quickest?

How many seedlings will you see?

How will you make your results reliable?

When warmed, a liquid evaporates to become a gas. When a gas is cooled, it becomes a liquid again.

Liquid —Evaporation→ Gas

Liquid ←Condensation— Gas

Getting the water back

Know that condensation is when a gas turns back into a liquid and is the reverse of evaporation.

The Big Idea

Condensation and evaporation are part of the water cycle and recycle the water on Earth.

Why does it rain?

- What would happen if all of the rain that falls stayed in the sea?
- What would happen if there was no liquid water on Earth?

When it rains, water falls from clouds and on to the ground or into the sea. The water on the ground can flow into rivers and then into lakes or the sea. The Sun then heats the water.

Air contains water vapour. If the air is cooled, the water vapour condenses. This water can fall as rain.

- Water can be heated by the Sun and move into the air as water vapour.
- Air can be cooled and the water vapour turns back to water.

As the water vapour cools, it forms clouds. When the clouds cool even more, the water will fall as rain. This is how the Earth recycles water. This is called the **water cycle**.

Amazing fact

One billion tonnes of water falls to the Earth every minute. That's about 130 trillion kilograms of water every day. That's a lot of water!

 Complete the diagram using the words in the word bank below.

Word Bank

Sun Evaporation River Sea Condensation Rain Clouds

Complete the paragraph using the words in the word bank below.

Water in rivers, lakes and __oceans__ is heated by the ______________ which makes it ______________. The water changes into ______________.

When this water vapour rises higher in the sky, it gets colder. This causes the water vapour to condense into liquid ______________. This is seen as ______________ in the sky. When these clouds cool even more, the water can fall to Earth as ______________. This process starts again and is called the ______________.

Word Bank

evaporate rain Sun water vapour water cycle clouds water ~~oceans~~

Now turn to page 58 to review and reflect on what you have learned.

Dry and damp air

Know that air contains water vapour and when this meets a cold surface it can condense.

The Big Idea

Air is made up of lots of gases, including water vapour.

Look at the photograph opposite. What can you see?

Why is this happening?

You may have noticed droplets of water forming on a cold glass.

 Where does the water come from?

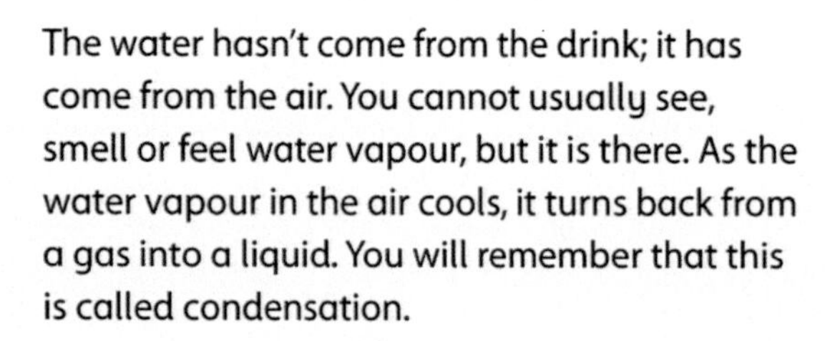

The water hasn't come from the drink; it has come from the air. You cannot usually see, smell or feel water vapour, but it is there. As the water vapour in the air cools, it turns back from a gas into a liquid. You will remember that this is called condensation.

In a very dry climate there isn't much water vapour in the air, but there is some. In wetter regions there can be a lot of water vapour in the air. The amount of water vapour in the air is called the humidity.

Did you know?

At sea level:

- air at 30°C can hold 28 grams of water vapour per cubic metre of air
- air at 10°C can hold only 8 grams of water vapour per cubic metre of air.

This means that warm air can hold more water vapour than cold air.

Place	Country	Rainfall per year (mm)
Arica	Chile	0.8
Lloro	Colombia	13 300
Aswan	Egypt	0.8
Kauai	Hawaii, USA	11 684
Al'Kufra	Libya	0.8
Mawsynram	India	11 811

Very dry and very wet places

 List the three driest places shown in the table above.

 Write the name of the wettest place shown in the table.

 Investigation: Measuring humidity

Hair changes length depending on whether it is wet or dry. Therefore, it is affected by humidity. In more humid or damp places, your hair will be longer.

Plan how you can use a strand of hair to measure the amount of water vapour in different places.

Think about:

- how you can safely obtain samples of hair
- how you will measure the hair before and after your investigation
- how you will compare dry and damp hair.

 Would your hair be longer in Arica, Aswan or Mawsynram?

Dry and damp air

Know that air contains water vapour and when this meets a cold surface it can condense.

The Big Idea

Water vapour can condense.

Water vapour moves around in the air very easily. This is why steam can reach the cool window at the other side of a room. If you heat a glass of water until it has all evaporated, the water vapour will fill the whole room.

Imagine that you overhear one of your friends explaining that the water on the window is because the window is leaking. How can you prove to them that the water is there because the water vapour in the air has condensed?

 Draw a diagram in your Investigation Notebook to explain where the water on the window has come from.

When we breathe, air moves into our lungs. It is warm and moist inside our lungs so the air is heated and picks up more water vapour. When we breathe out, the water vapour spreads into the air. When it is cold, some of this water vapour turns into water droplets and these droplets can be seen. So when it is very cold and we breathe out, we can see a cloud. A cloud is made up of billions of droplets which fall as rain when they get heavy enough.

You can try this by wetting your finger and then blowing on to it.

Write down what you feel and then explain what is happening to the water on your finger.

Why does a cloud appear when we breathe out in a cold place?

Why don't we see this if the place is warm?

What is the scientific name for water vapour changing into water droplets?

Condensation is a big problem in countries that are damp and cold. Water vapour can cool and cause dampness in houses. This can make wood and other building materials rot away. To help, people keep their houses warm and allow air to flow through them. This is called ventilation.

We can use evaporation to help us keep cool. For example, when runners are hot they sweat. The moisture on their skin then evaporates into the air. This uses a lot of energy and cools the runners down.

Think about...

Keeping a house warm and well ventilated makes condensation less of a problem. Why?

Decide whether the following statements are true or false.

		True/False
1	Air contains water vapour.	________
2	Humidity affects the length of your hair.	________
3	We do not breathe out water vapour.	________
4	The water leaks out of the glass of iced water on to the outside of the glass.	________

Now turn to page 59 to review and reflect on what you have learned.

Boiling and melting

Know that the boiling temperature of water is 100°C and the melting point of ice is 0°C.

The Big Idea

Every liquid has its own special boiling point and melting point.

Sometimes we want to have water that is very hot. The equipment above is being heated to make it safe to use in operations. The hot water kills microorganisms.

- Where have you seen or used very hot water?
- Why must we be very careful with hot water?
- How do you know that water is very hot just by looking at it?

Temperature tells us how hot or cold things are. To measure temperature, we use a thermometer. We measure temperature in degrees Celsius. This is written as °C.

- Label the diagram of the thermometer by completing each box with the correct common substance from the word bank below.

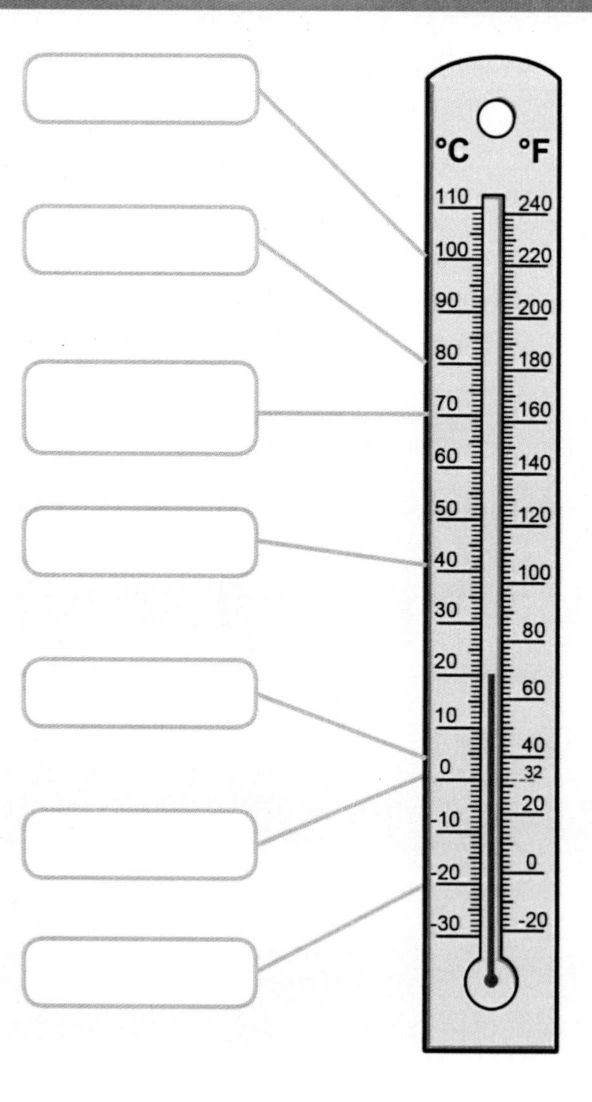

Word Bank

Hot coffee	Bath water	Freezer	Ice
Fridge	Hot tap water	Boiling water	

Practise using a thermometer by finding out the temperature of:

- the room
- water from the cold tap
- your skin.

When a liquid is heated enough, it boils. The temperature at which a liquid boils is called its **boiling point**. When things reach boiling point, they change from a liquid to a gas. The boiling point is the highest temperature a liquid can reach unless it is in a pressure cooker.

The boiling point of water is 100°C. At this temperature, the liquid water changes into very hot water vapour called steam. When water is boiling, bubbles are formed throughout the liquid but the biggest ones are at the top.

Amazing fact

The surface of the planet Venus is 480°C. This is nearly five times hotter than boiling water. If this is too hot for you, why not move to the planet Neptune? Here, the temperature is a cool −200°C!

Boiling and melting

Know that the boiling temperature of water is 100°C and the melting point of ice is 0°C.

The Big Idea

Boiling points and freezing points can be measured.

Investigation: What is the boiling point of water?

Plan an investigation to prove that the boiling point of water is 100°C.

You must:

- write down a plan and include a diagram
- list any safety factors to think about
- decide how you will heat up your water
- make sure you take accurate readings on your thermometer.

Be very careful with boiling water. It can scald you.

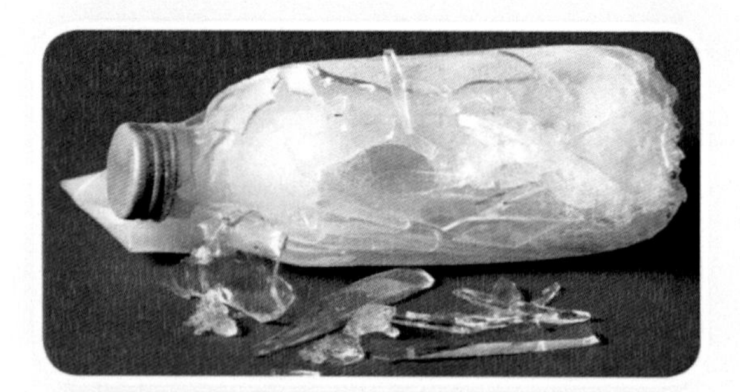

When a liquid cools, it changes from a liquid into a solid. This is called freezing.

Each different liquid will freeze at a certain temperature called its freezing point. Water freezes at 0°C. You will remember that solid water is called ice. When water freezes, it increases in size. This is why a full glass bottle of water might break if you put it in a freezer.

Why are frozen pipes a problem in very cold countries?

If you heat a solid to a high enough temperature, it will begin to melt. Each different solid will turn into a liquid at a temperature called its **melting point**.

A clue about melting

List four things that you have seen melt.

The boy in the picture is upset. His ice-lolly seems to be disappearing.

What is really happening to the ice-lolly?

What advice can you give the boy to help him keep his ice-lolly for longer?

Substance	Melting point (°C)	Boiling point (°C)
Oxygen	–218	–183
Water	0	100
Chocolate	30	180
Iron	1535	2750
Salt	808	1465
Copper	1086	1187
Candle wax	65	360
Diesel	–19	154

Look at the table above.

Which two substances are liquids at 25°C?

Which five substances are solids at 25°C?

Which substance would have to be cooled to nearly –220°C to make it a liquid?

Not all the same

Different substances have different boiling points and melting points. This is vital, otherwise we wouldn't have the three states of matter on Earth. Everything would be in the same state.

Think about...

How can you check that the boiling point of water is lower on mountains?

Amazing fact

Water only boils at 100°C at sea level. On mountains, it will boil at lower temperatures as the air is thinner. On the highest mountain in the world, water boils at 71°C.

Boiling and melting

Know that the boiling temperature of water is 100°C and the melting point of ice is 0°C.

The Big Idea

Scientists use results to make decisions.

 Explain the term 'melting point'.

 Explain the term 'boiling point'.

The particles in solids, liquids and gases are arranged differently.

 Look back to your previous work on particle diagrams on page 28. In your Investigation Notebook, draw the particle diagram of water and the particle diagram of copper.

The way that the particles are arranged affects the melting, freezing and boiling points of substances.

Liquid	Boiling point (°C)
Olive oil	300
Petrol	210
Jet fuel	163
Paraffin	150
Vinegar	118
Sea water	108
Water	100

The table shows the boiling points of different liquids. Tables are a good way of recording your results.

 What pattern in the results can you see from the table?

It can be difficult to see patterns in a table and to write a conclusion about the results. Scientists and mathematicians often display results like this in charts.

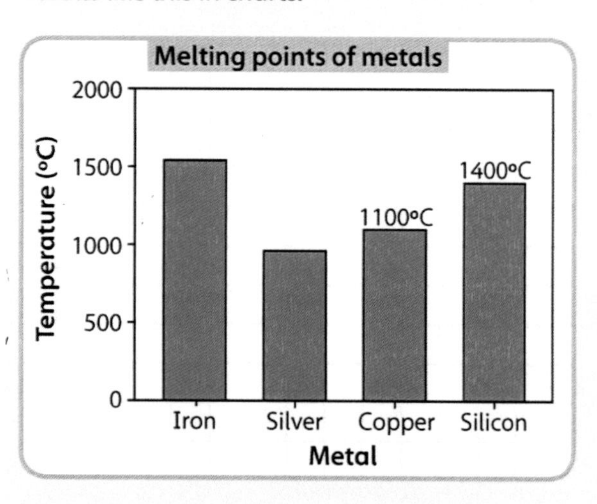

This bar chart shows the melting points of some metals. It is much easier to see a pattern in this chart and you can clearly see the highest and lowest melting points.

 Look at the bar chart. Which metal has the highest melting point?

 Which metal has the lowest melting point?

 What is the melting point of copper? (Don't forget to write the unit it is measured in.)

 What is the melting point of silicon?

Only water that has nothing added to it boils at 100ºC. Scientists call this pure water. We can test if a liquid is pure water by measuring its boiling point.

 Investigation: Boiling temperatures of water samples

Imagine you are a scientist trying to find out if a sample is pure water.

This table of results is from an investigation carried out by chemists. The chemists investigated three samples of water and measured the boiling temperature of each sample three times. They have rounded the averages to whole numbers. Plot their average results on a bar chart. Make sure you follow the graph-drawing rules below:

- Label each axis and include the units, for example °C.
- Make your scale even.
- Plot the data accurately.
- Use a sharp pencil and a ruler.

 Write a concluding sentence based on the results shown in your bar chart.

Sample	Boiling temperature (ºC)	Boiling temperature (ºC)	Boiling temperature (ºC)	Average boiling temperature (ºC)
A	100	100	99	100
B	93	94	94	94
C	110	112	110	111

Boiling and melting

Know that the boiling temperature of water is 100°C and the melting point of ice is 0°C.

The Big Idea

Scientists control variables when they plan investigations.

 Write two reasons why clean water is so important to life on Earth.

Where does your water supply come from? How do you look after it?

Look at the three glasses of water above. Which one would you drink?

Discuss the reasons for your choice. Why didn't you like the other glasses of water?

It is not really possible to decide just by looking. This is because we cannot see many of the things in water that make us ill. They are microscopic. This means that they can only be seen with a powerful microscope.

The factors that impact on an investigation are called variables. For example:

- the amount of material you will use
- the temperature you use
- the amount of gas given off.

The variable you are going to deliberately alter is called the independent variable.

The variable that will be measured is called the dependent variable.

All of the other factors in an investigation that scientists try to keep the same are called the control variables.

Scientists use lots of different ways to find out how pure water is.

Investigation: The purity of water samples

You are going to find out which of the three water samples you have been given is the purest.

⚠ Be careful when heating your samples as they will get very hot.

How will you know that your sample is boiling? What will you observe?

Predict which one of the samples will be the purest.

In your Investigation Notebook, copy and complete the table below.

Sample	A	B	C
Boiling temperature (ºC) Try 1			
Boiling temperature (ºC) Try 2			
Boiling temperature (ºC) Try 3			
Average boiling temperature (ºC)			

Remember

Only pure water boils at 100ºC.

This investigation proves that even though the water looks pure, it may not be. To get accurate results, a scientist will have to carry out the same test many times. This will give repeat measurements.

Which control variables did you keep the same?

Is there anything that you would change if you did this test again?

1 Why do scientists use graphs and charts?

2 What is the boiling point of water?

3 What is the melting point of ice?

Now turn to page 59 to review and reflect on what you have learned.

Making solutions

Know that when a liquid evaporates from a solution a solid is left behind.

The Big Idea

A solution is made by dissolving a solid in a liquid.

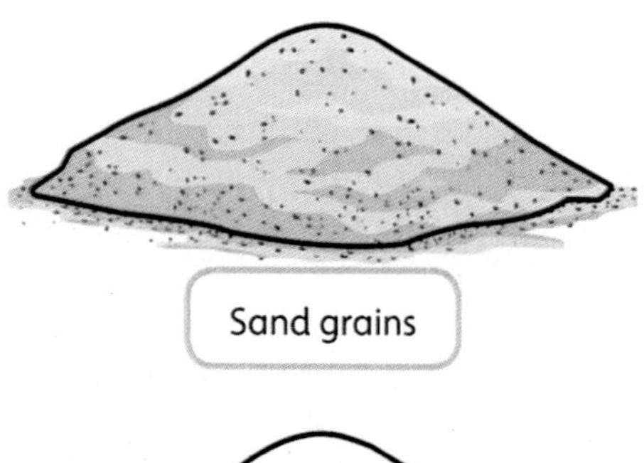

Sand grains

Salt grains

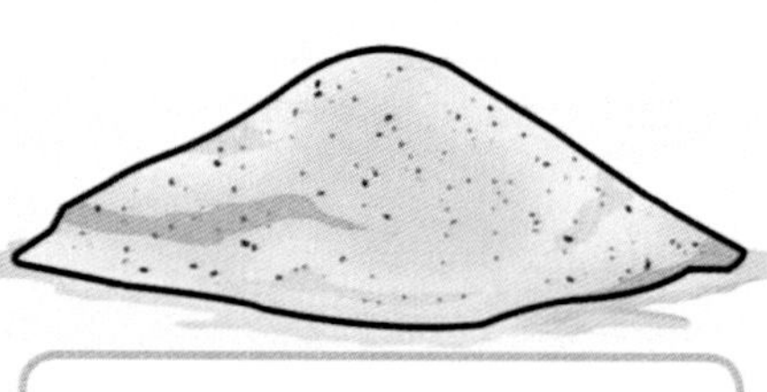
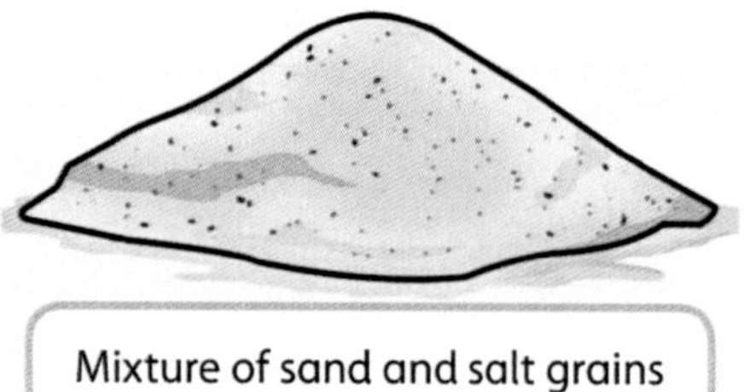

Mixture of sand and salt grains

Look at the pictures of sand and salt grains above. Where have you seen these substances before?

Predict what will happen if the sand and salt mixture is added to water and stirred.

Solids, liquids and gases can be mixed together. Some solids dissolve in water. This means the solid cannot be seen any more.

The soild has not disappeared but has mixed into the water. A liquid with a dissolved solid in it is called a **solution**. There are lots of examples of solutions, such as salt water.

Is sea water a solution? Why?

We use solutions in our everyday life. Coffee is a solution. It is hot water with chemicals from the coffee beans dissolved into it. Medicines such as aspirin can be dissolved in water to make them easier to swallow. Fruit juices are mixed with water to make a solution.

Write four examples of solutions that you have seen.

 Investigation: Dissolving

1 We know that salt dissolves in water. Investigate which other substances dissolve in water.

2 Measure 50 cm^3 of water into a glass beaker. Add one teaspoon of salt and stir your mixture. Then add another teaspoon of salt and stir the mixture again. Keep adding salt until no more will dissolve. You will know when no more salt will dissolve because solid salt will stay in the beaker.

3 Now repeat this investigation using other solids.

 Why should you use the same amount of water every time?

 What else must you keep the same in order to make this a fair test?

 In your Investigation Notebook, design a table to record your results.

 Which substances were the best at dissolving?

If a substance dissolves well in water, we say that it is very **soluble**. If a substance does not dissolve in water, we say that it is **insoluble**. You may have found some insoluble substances in your investigation. Sand is insoluble.

Why is it important we have both soluble and insoluble materials on Earth?

What would happen if metals were soluble in water?

What happens to coffee powder when you add it to hot water?

Making solutions

Know that when a liquid evaporates from a solution a solid is left behind.

The Big Idea

Crystals can be made from a solution.

What is a solution?

Why are solutions so important? Think of two examples that you have seen or used today.

If you make a concentrated solution, you can make **crystals**.

Investigation: Making crystals

1 Heat some water in a beaker. Put one teaspoon of salt into the liquid. Stir in the salt until it dissolves. When the salt has dissolved, you will no longer be able to see it. It has dissolved into the liquid. Keep adding more and more teaspoons of salt until the solution is saturated. This means that the liquid cannot dissolve any more salt.

2 Pour some of the solution into an evaporating dish or a shallow dish. Put it somewhere warm. For example, you might use a sunny windowsill. Leave the dish for as long as you can. Keep looking at it and you should see crystals forming.

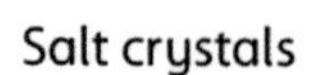

Salt crystals

Copper crystals

It is possible to make crystals using lots of different solids, such as sugar. Lots of chemicals also make colourful crystals.

Investigation: Making a crystal garden

1 Sprinkle salt on to a tray. Carefully pour some of the solution that you have on to the tray and scatter a few rocks.

2 Leave the tray somewhere warm where it cannot be disturbed. After some time, you will have a crystal garden.

3 Have a competition to see who can grow the biggest crystal. To do this, you need to carry out a fair test.

Natural crystals in Cueva de los Cristales, Mexico

The longer the crystals take to form the bigger they are. This means that the crystals above have been forming for many thousands of years. Water containing dissolved minerals drips into caves. As this happens, the water evaporates leaving stalagmites or stalactites. Stalagmites rise up from the ground. Stalactites hang down from the top of caves.

Investigation: Making stalagmites

Try making a stalagmite for your crystal garden.

1 Make up a saturated solution of salt water in a beaker.

2 Tie a piece of cotton to a pencil and rest the pencil on top of the beaker containing the solution. Make sure the cotton or string is just touching the surface of the solution.

3 Leave this for a few days, making sure your solution does not evaporate completely. Keep adding more solution if it is very hot in your classroom.

4 When you have formed a stalagmite on your cotton or string, you can put it in your crystal garden.

Stalactites

Explain what happens to the water when you make crystals.

Making solutions

Know that when a liquid evaporates from a solution a solid is left behind.

The Big Idea

Sieving, filtering and evaporation can be used to separate mixtures.

 List three substances that dissolve in water.

 List three substances that do not dissolve in water.

Remember, substances that dissolve in water are soluble. Salt and sugar are good examples.

You found out that not everything dissolves in water. Sand, rocks, clay, chalk and oil do not dissolve in water. These are called insoluble substances.

If you mix insoluble materials with water, most of them just sink to the bottom. The rest of them float around in the water but they do not dissolve.

 How can you separate the insoluble solids from the water?

How can you then separate the solids into different piles?

Removing the water

We know that we can remove water by heating. The water will evaporate. This is useful, but not if we want to keep the water. A simpler way to separate the solids from the water is to use **filter paper**.

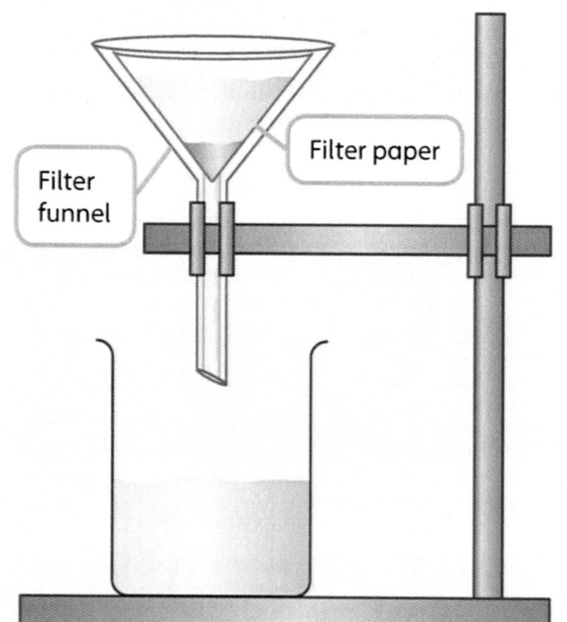

The filter paper is full of very small holes. Water runs through the holes and is caught in the beaker. The insoluble solids cannot pass through the holes. They are caught in the **filter funnel**. This method is called filtration.

How does filter paper catch solids?

What is this process called?

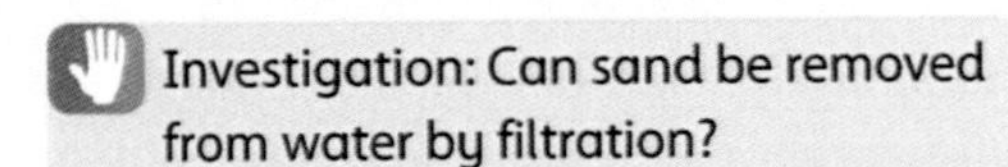

Investigation: Can sand be removed from water by filtration?

Plan an investigation to separate sand from water. Use similar equipment to that shown in the picture opposite.

When salt dissolves in water, the salt spreads out as tiny particles or pieces. Each particle is very small and can pass through the holes in the filter paper. Dissolved substances cannot be separated by filtering them.

Draw lines to match up the heads and tails of the sentences below.

Dissolved substances cannot be separated by …	… filter paper.
When salt spreads out in water we say it has …	… filtration.
Sand does not dissolve. It can be trapped in …	… dissolved.

The larger insoluble materials can be separated by sieving. Cooks and bakers use **sieves** to separate lumps from flour. Some people use a sieve called a tea strainer to separate the tea leaves from a cup of tea.

Describe when you have seen a sieve being used.

Sieves can have different sized holes to help separate different sized particles. They can be used to divide soil into small, medium and large pieces.

How can you use different sized sieves to separate the sand, pebbles, cork and paper clips shown in the beaker opposite?

The paper clips are made of steel. Can you think of another way to separate them from the mixture?

Making solutions

Know that when a liquid evaporates from a solution a solid is left behind.

The Big Idea

Science can solve problems.

What does an environmental scientist do?

A complaint has been made by a chef in a local restaurant. He bought a bag of rice, but there seems to be something else in there. The chef is puzzled about this. He does not want to use the rice until he finds out what else is in the bag. He contacts a local environmental scientist to help him find out what is in the bag.

There seem to be lots of different ingredients and foods in the bag. Can you help him to separate them?

Investigation: Separating the mixture

Look at the sample you have been given. It looks like someone has mixed up salt, sugar, sand, rice, dried peas and paper clips. These have all been mixed with water.

You will have to take the sample back to your laboratory where all your equipment is. Some pieces of equipment that you could use are shown below.

Write a step-by-step plan to separate all of the things in the mixture. Use the following prompts to guide you:

- What will you separate first?
- What equipment will you use to do this?
- What will you use the magnet for?
- What will you use to separate the sand?
- How will you get the salt and sugar from the solution?

Salt and sugar look very similar. If you look carefully at the crystals, the sugar is a different shape. Salt crystals have six sides and look like a cube. Sugar crystals are six-sided too, but are stretched into longer shapes. Sugar crystals are also brighter than salt crystals. The salt looks a dull grey next to the sparkly sugar.

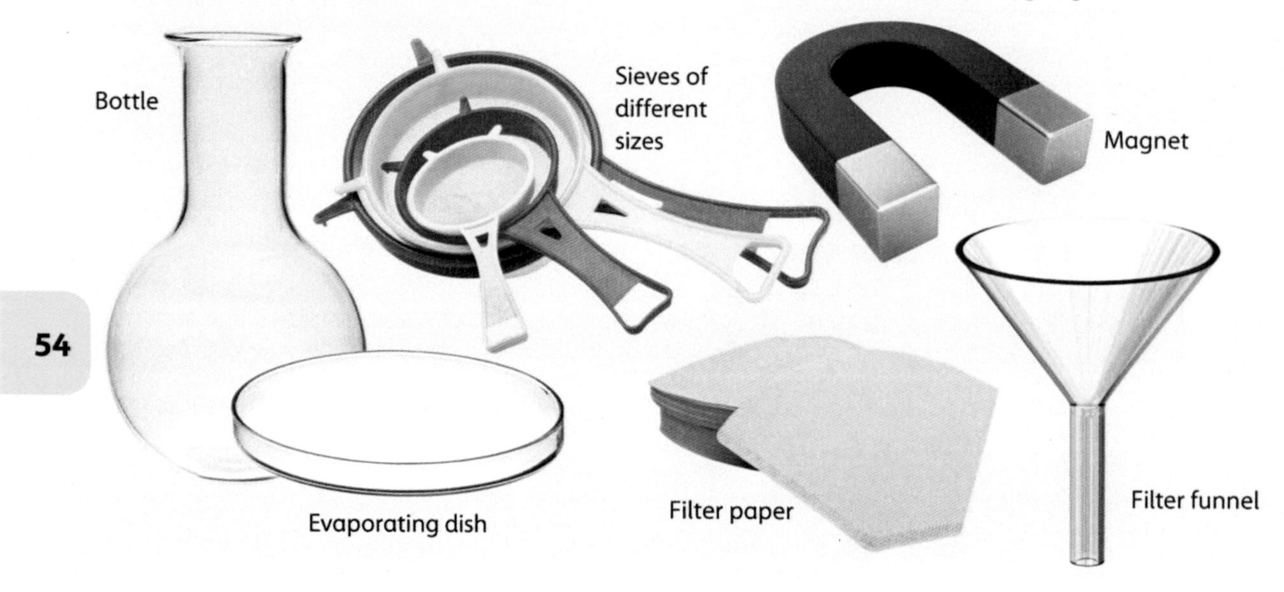

Salt dissolves in water more easily than sugar. You might be able to use this information to help you to separate them.

A slow way to separate salt and sugar would be to use a hand lens or microscope and pick out the different shaped crystals. You might be able to use this information to help you separate them.

Scientists use special liquids to separate salt and sugar. They use liquids that let sugar dissolve but not salt. These are special solvents. A solvent is a liquid that will dissolve a solid or a gas. Some of these are used to clean paint brushes and remove spilled oil.

Salt crystals

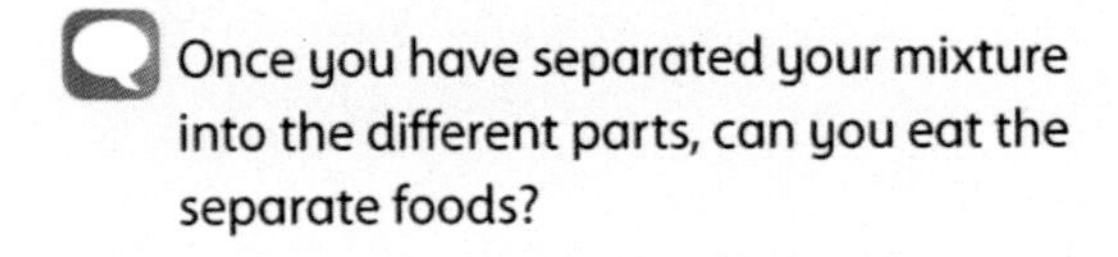
Once you have separated your mixture into the different parts, can you eat the separate foods?

When would you use evaporation? Circle the correct answer.

To separate a metal from a mixture.

To separate larger particles.

To separate dissolved substances.

Decide whether the following statements are true or false.

		True/False
1	It is safe to eat food during an experiment.	________
2	Magnets can be used to separate sugar from salt.	________
3	Sieves can be used to separate larger solids.	________

Sugar crystals

Making solutions

Know that when a liquid evaporates from a solution a solid is left behind.

The Big Idea

We can measure the amount of a substance dissolved in water.

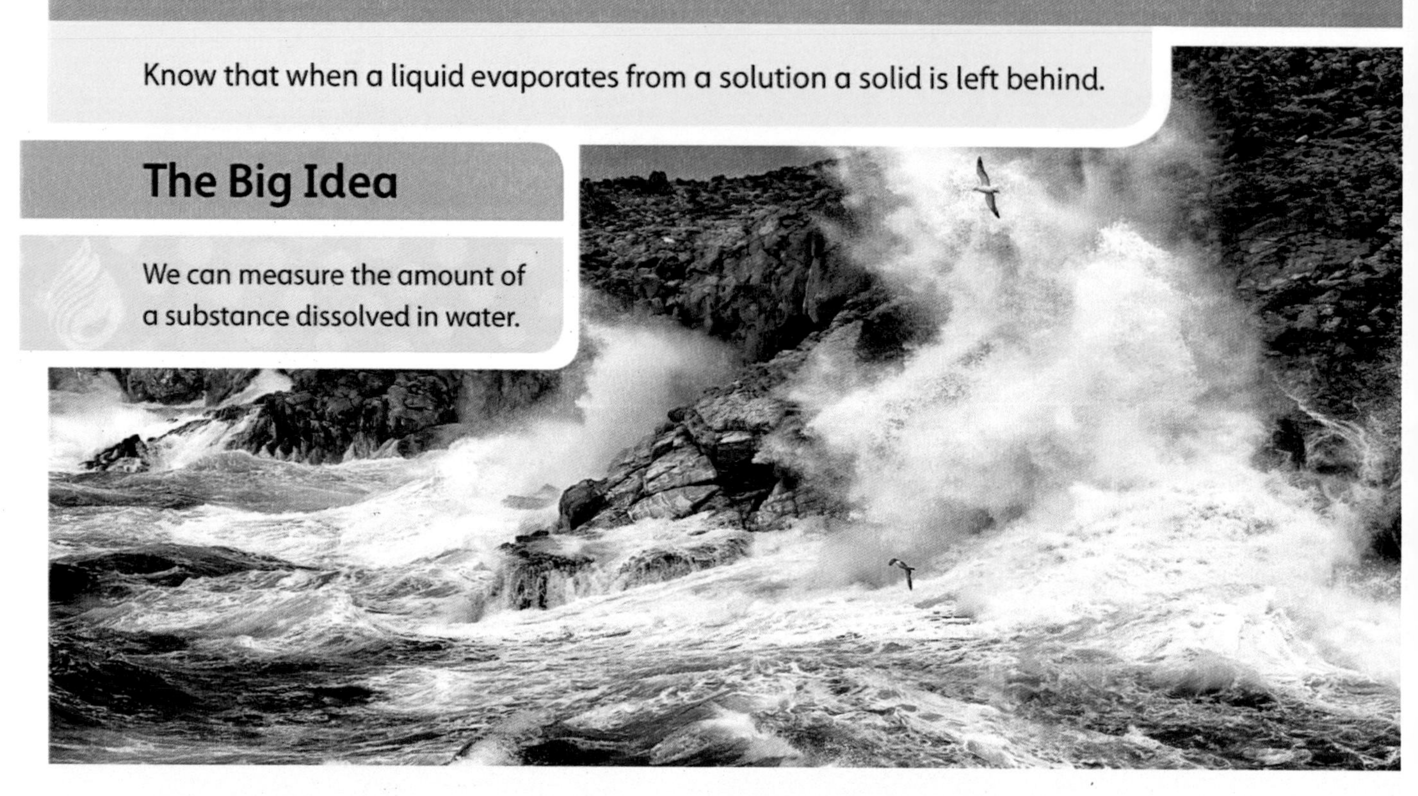

Sea water is a solution.

If you swim in the sea and then dry in the Sun, you can see crystals on your skin. This is the salt that was dissolved in the sea water.

 Where does the water go?

It is estimated that 70 per cent of the Earth's surface is water. Most of this is in the seas and oceans. That is a lot of water. In many countries around the world people suffer from droughts, which means there is not enough water to drink or water crops.

 If you leave a dish of sea water near a source of heat, the water will …

Salt panning

 What will be left in the dish?

Amazing fact

1260 million trillion litres of water make up the oceans and seas on Earth.

Investigation: Testing saline solutions

In hospital, people are sometimes given a saline drip. Saline is a solution of salt in water.

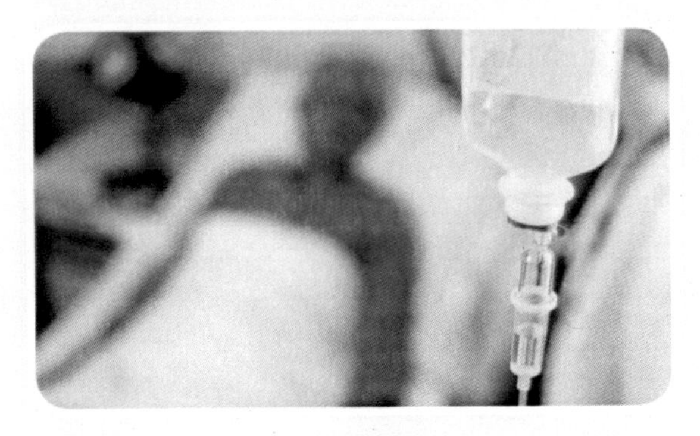

It is very important that the saline is not too salty. The solutions used in hospitals have 9 grams of salt in every 1000 cm^3 (litre) of water. If more salt than this is added, it may make the person even more ill.

One hospital has a problem. The labels on the salt solutions have fallen off. You need to find out which of the three solutions is the one to give the patient.

The three saline solutions are A, B and C.

Plan your investigation. Think about:

- how you will remove the water from the saline solution
- how you will find out how many grams of salt are left behind
- how you will make this a fair test.

 If you are only given 100 cm^3 of water, how many grams of salt should the correct saline solution contain?

Think about...

Find out how water can be purified in water treatment plants using large-scale filters. How is it similar to how water is cleaned in nature? Will this type of filtering remove substances that are dissolved in water?

 1 How can you separate chalk from chalky water?

2 What is a substance that dissolves in water called?

3 What is a substance that does not dissolve in water called?

4 How can you speed up the process of dissolving?

5 What process can you use to separate salt from a solution?

Now turn to page 59 to review and reflect on what you have learned.

What we have learned about evaporation and condensation

Where does the water go?

(pages 24–31)

I know that when a liquid becomes a gas it is called evaporation.

I know that when water turns into a gas it is called water vapour.

I can model the particles in solids, liquids and gases.

At what temperature does water boil?

Liquid water from wet clothes evaporates. What does this mean?

What are the three states of matter?

Getting the water back

(pages 32–5)

Why does steam condense back into water when it lands on some surfaces?

What are the two processes that turn water into clouds and clouds into rain?

I know that condensation is the reverse of evaporation.

I understand the importance of the water cycle on Earth.

Dry and damp air (pages 36–9)

Decide whether the following statements are true or false.

		True/False
1	Air contains water vapour.	________
2	We do not breathe out water vapour.	________
3	Water leaks out of a glass of iced water on to the outside of the glass.	________

I know that humid air contains water vapour.

Boiling and melting (pages 40–7)

What happens when a liquid reaches its boiling point?

What happens when a liquid reaches its freezing point?

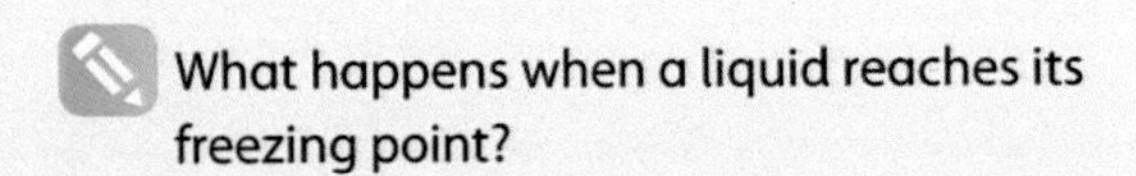

What happens when a solid reaches its melting point?

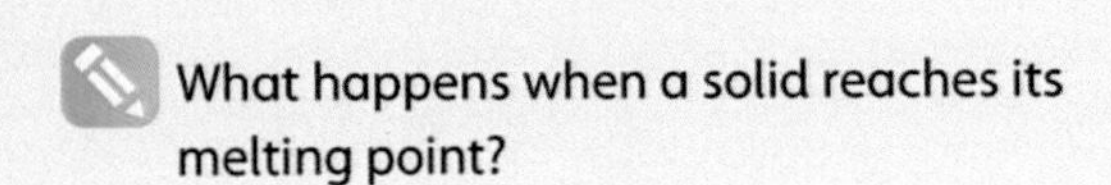

I know that different liquids have different boiling, freezing and melting points.

I know what an independent variable and a dependent variable are.

I understand that to make a fair test the other variables have to be controlled.

Making solutions (pages 48–57)

When solids disappear in a liquid they have …

Name two solutions made with water and solids.

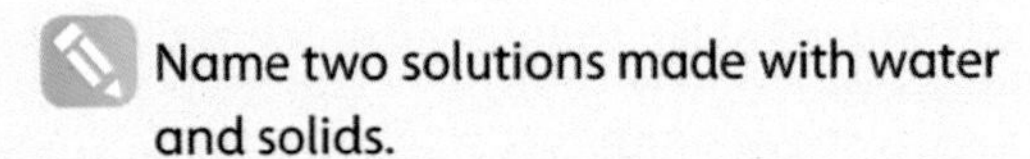

Name one solid that is insoluble in water.

I know that some substances are soluble and some are insoluble.

I know that when a liquid is evaporated from a solution a solid is left behind.

3 The Life Cycle of a Flowering Plant

In this module you will:

- understand that plants reproduce
- learn about how seeds can be dispersed
- discover that insects pollinate some flowers
- learn that plants produce flowers that have male and female parts
- find out that seeds are formed when pollen from the male part fertilises the ovum (female)
- learn that flowering plants have a life cycle which includes pollination, fertilisation, seed production, seed dispersal and germination.

Amazing fact

Did you know that a watermelon may have over 600 seeds inside?

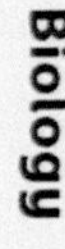

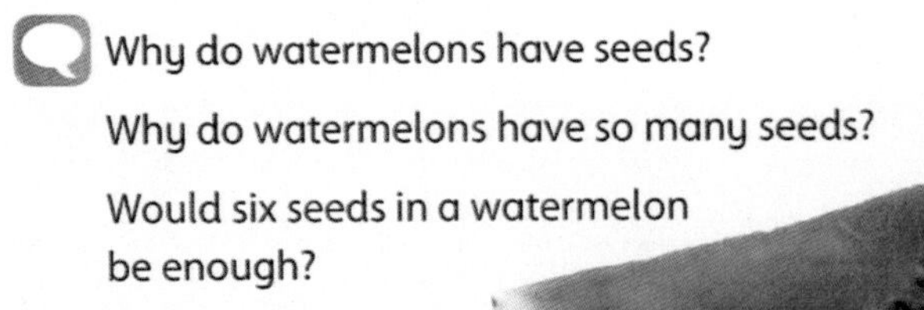

Why do watermelons have seeds?

Why do watermelons have so many seeds?

Would six seeds in a watermelon be enough?

Word Cloud

seed production
explosion
insect
fertilisation
dispersal
male (anther)
female (ovum)
germination
wind
interpret findings
life cycle
fruit
pollination
present

Flowering plants

Know that plants reproduce.

The Big Idea

Flowering plants reproduce. They start as a seed and grow into adult plants.

How do flowering plants make new flowering plants?

When flowering plants reproduce they make new plants. These new plants grow and reproduce to make more new plants. We call this process the **life cycle** because each stage in the life cycle is repeated each time a new plant is made. There are four main stages in the life cycle of a flowering plant.

1 Seeds

We can think of seeds as the first stage in the life cycle of flowering plants. Seeds need to have the right conditions, such as water and warmth, to start to grow. Once a seed starts to grow we say it has **germinated**.

2 Seedlings

We can think of seedlings as the second stage in the life cycle. We start to see the first shoots and roots at this stage.

3 Young plants

We can think of young plants as the third stage in the life cycle. At this stage we should be able to identify what kind of plant it is.

4 Adult plants

We can think of adult plants as the fourth stage in the life cycle. At this stage the plant is fully grown.

Look at the photographs of different plants below. Write the name and number of each stage in the life cycle to complete the table. The first one has been done for you.

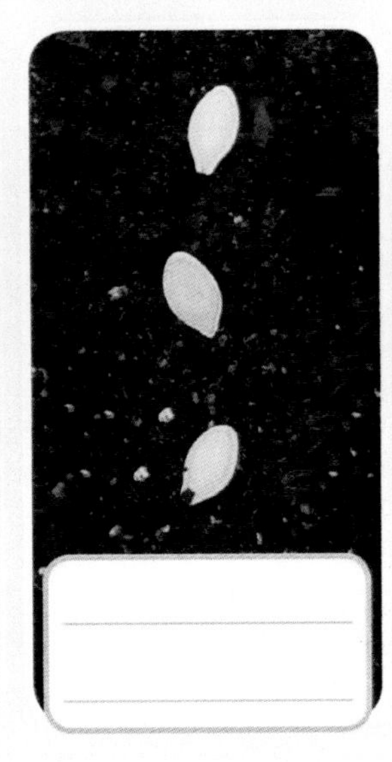

Circle the correct words. The first one has been done for you.

The life cycle/circle of a flowering plant always/never starts with one plant. When a flowering plant reproduces/produces it can make seeds. The first stage in the life cycle of a flowering plant is a seed/young plant. As the seed grows it moves into stage two of the life cycle to become a seedling/adult plant. Stage three of the life cycle is the young plant/seed and stage four is the adult plant/seed. When seeds start to grow we say they germinate/reproduce.

Think about...

1. The watermelon plant hides its seeds inside a fleshy **fruit** that is good to eat. Why?
2. What happens to the seeds when you eat the fruit?
3. Why does the watermelon produce 600 seeds? Isn't it easier to produce six seeds?

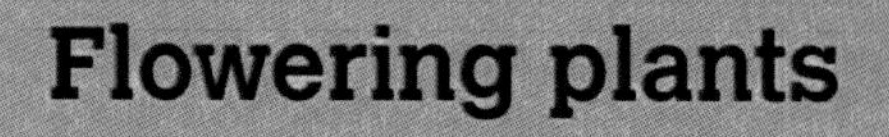

Flowering plants

Know that plants reproduce.

The Big Idea

We can show the life cycle of flowering plants as a circle.

At which stage of the life cycle do we first see roots and shoots?

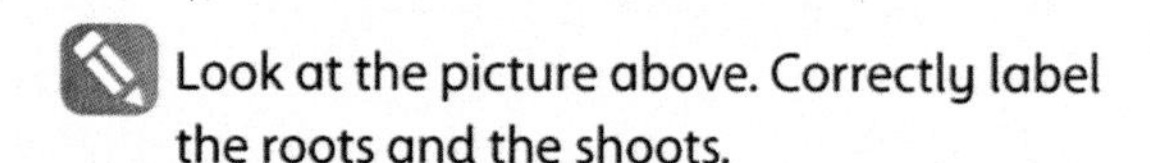

Look at the picture above. Correctly label the roots and the shoots.

We have looked at the life cycle of a flowering plant and explored some of the stages. The life cycle of a flowering plant is often shown as a circle. This shows us that the life cycle is a continuous process – it never ends.

Look at the picture on the right. It shows a simple life cycle of a flowering plant. Label the life cycle using the words in the word bank below. One has been done for you.

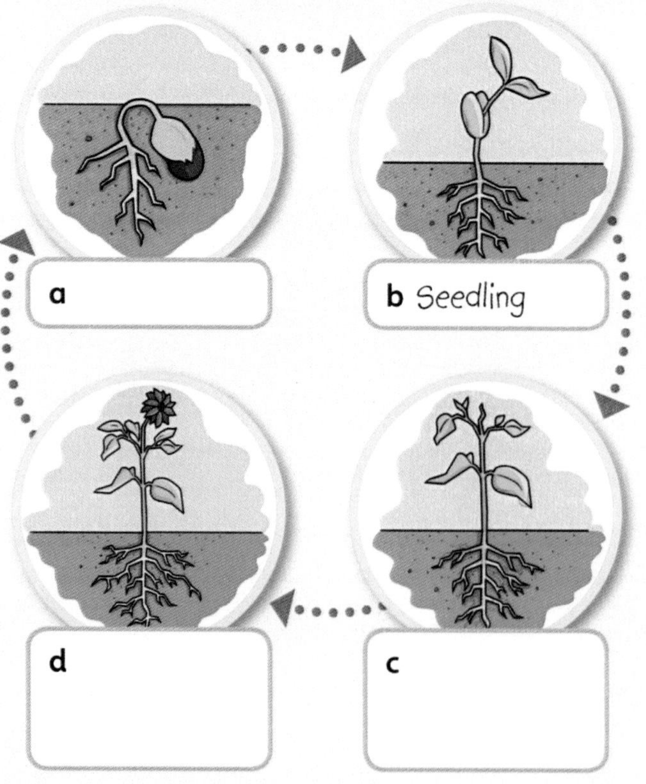

Word Bank

Seed	Young plant
~~Seedling~~	Adult plant

But what about flowering plants that produce fruits? There needs to be another stage in the life cycle diagram that shows the plant producing fruit.

What is your favourite fruit? Use the space below to create a life cycle of your favourite fruit showing the plant producing the fruit as an additional stage. Label your life cycle using the labels in the word bank below.

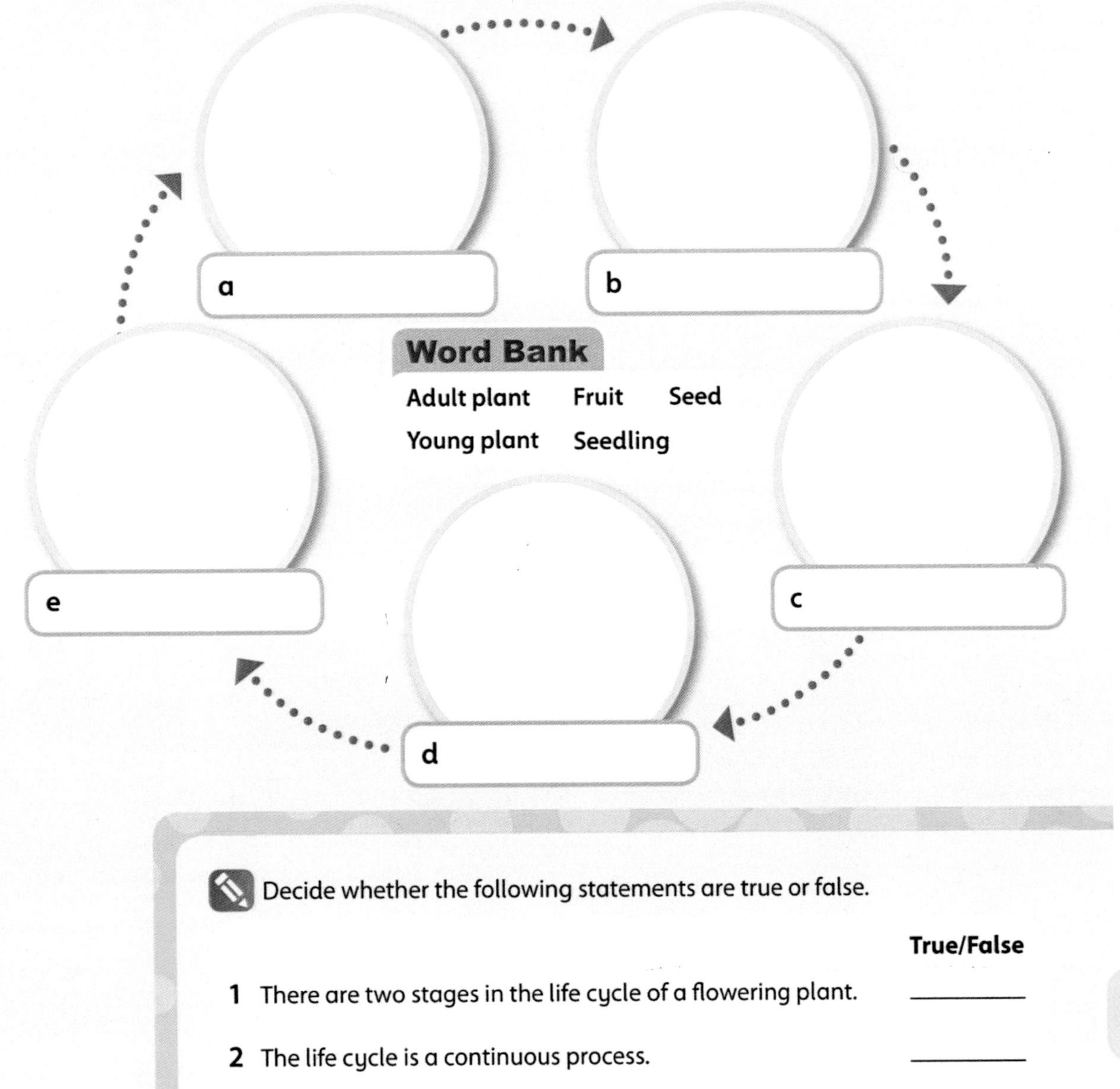

Decide whether the following statements are true or false.

		True/False
1	There are two stages in the life cycle of a flowering plant.	________
2	The life cycle is a continuous process.	________
3	Some flowering plants produce fruit with seeds in them.	________

Now turn to page 82 to review and reflect on what you have learned.

Seeds, seeds everywhere!

Learn about how seeds can be dispersed.

The Big Idea

Seeds need to find a place to grow.

Do you remember what seeds need to start to germinate?

What happens to the seeds?

We know that flowering plants need to produce seeds to make new plants, but what happens to the seeds?

To find somewhere to grow with enough space, water and soil, seeds need to move away from where they are made. When seeds have been spread out, we say they have been **dispersed**.

Not every seed can find somewhere to grow into a new plant and so not all seeds germinate. Plants produce a lot of seeds so that some of them will find a place to grow.

There are four main ways that seeds are dispersed:

1 Wind

Usually the seeds are very lightweight and can be blown by the **wind**. When the wind drops them, they can then grow into new plants.

2 Explosion

Some plants can disperse seeds by **explosions**. Their seed pods actually shoot the seeds out. The seeds can then grow into new flowering plants.

3 Animals and birds

Seeds can be carried a long way by animals and birds. Some seeds stick to their fur, are eaten and then pass through the animal unharmed or are buried for food and forgotten about.

4 Water

Some plants that live in or near water can use water to help disperse their seeds. Seeds can travel a long way on water.

 Look closely at the four photographs below. Write the names of how the seeds are dispersed in the boxes. One has been done for you.

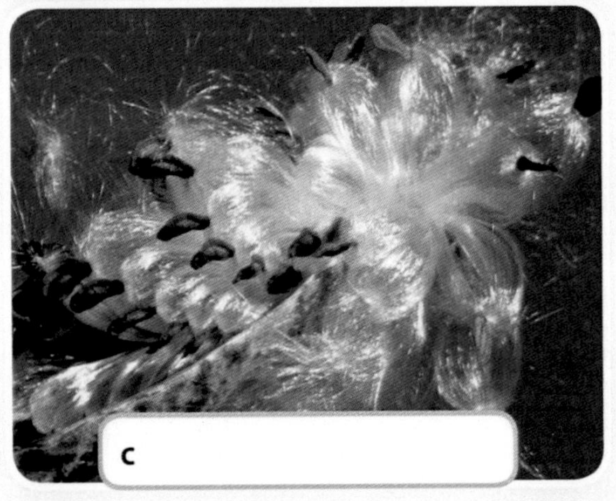

When the seeds have been dispersed they start to grow. We already know that the seeds need the right conditions, like water and warmth, to help them to grow. When the seeds find the right conditions to grow, they start to germinate.

Complete the paragraph using the words in the word bank. One has been done for you.

Seeds need to be ______________ to find a place to grow. There are ___four___ main ways of seed dispersal. They are ______________, explosion, ______________, animals and birds. When the seeds find a place to grow they start to ______________.

Word Bank

~~four~~ water dispersed wind germinate

Seeds, seeds everywhere!

Learn about how seeds can be dispersed.

The Big Idea

Seeds sometimes have competitions to find the right place to grow!

List the four main methods of seed dispersal.

Seeds come in a wide range of shapes and sizes. The purpose of seeds is to germinate and grow into new plants. Sometimes there is not enough room for seeds to grow, which is why they need to be dispersed. Sometimes there is a lot of competition for space, so some seeds need to travel a long way before they can find the right conditions to germinate.

Seeds can be adapted to get a long way away from the parent plant. Let's look at some of these adaptations.

Look at the photographs. Discuss with a partner what the seeds look like and then write which method of seed dispersal is best for each type of seed. The first one has been done for you.

Complete this paragraph about adaptations of seeds using the words in the word bank below. One has been done for you.

Seeds sometimes have to compete to find a place to grow. Seeds can ______________ so they can move far away from the parent plant. Seeds that are dispersed by the wind are usually very ______________. Seeds that are dispersed by water are usually able to ______________. Seeds that are dispersed by animals may have parts that ______________ to the animal's fur or may have tasty fruits that the animals ______________. Seeds that are dispersed by explosions usually ______________ the seed pod and are then carried by the wind.

Word Bank

float **lightweight** ~~**water**~~ **adapt** **stick** **burst** **eat**

Now turn to page 82 to review and reflect on what you have learned.

Pollinating flowers

Discover that insects pollinate some flowers.

The Big Idea

Flowering plants need to be pollinated so they can produce seeds.

Why do insects like flowering plants?

Some flowering plants need help from **insects** to produce seeds. Some insects move around from flower to flower. When insects do this they are also helping the plant to produce seeds.

 Look closely at the photograph above. What is the insect doing?

If you look very closely, you can see that the insect has small particles on its body. These particles are called pollen.

Pollen is very important because it is used by the plant to make seeds. As the insect flies or moves from plant to plant, pollen from one plant sticks to the insect and is dropped on to another plant. When insects drop pollen from one plant on to another plant of the *same* kind, we call this **pollination**. We say that the insect pollinates the plant.

Some flowering plants produce very sweet liquid called nectar. Nectar provides food for many insects. The insects and the plants help each other because the plant produces food for the insect and the insect pollinates the plant.

Are there other ways to pollinate plants?

We now know that insects have a big role to play in pollinating plants, but they are not alone! Plants can be pollinated in other ways too.

Plants can be pollinated by humans

Pollen can travel on water to other plants

Birds can help to pollinate plants

Why do humans want to help pollinate plants?

List two different ways in which pollen can move from one plant to another plant of the same kind.

Complete the paragraph using the words in the word bank below. One has been done for you.

Insects use flowering plants for __________. The food produced by some flowering plants is very sweet and is called __________. When the insect is feeding, small particles called pollen stick to its body. As the insect moves from one plant to another plant of the same kind, it can drop pollen. When insects do this we call it __________. Pollination can be done by birds, animals, __________, water and wind.

Word Bank

humans nectar ~~pollen~~ food pollination

Pollinating flowers

Discover that insects pollinate some flowers.

The Big Idea

We can investigate that bees pollinate flowers.

 Investigation: Do bees prefer petals or nectar?

Part 1: Petal colour

Scientists conducted an investigation to find out how flowers attract insects and other pollinators. First, they thought about the colour of the petals.

They studied bees to find out if they preferred one petal colour to another. The scientists used artificial flowers of different colours. To make it a fair test they made sure the petals were the same size and that observations were made at the same time of the day.

The scientists placed the flowers where there were lots of bees and then counted how many times the bees visited the flowers.

Petal colour	Number of visits by bees
Blue	46
Red	15
White	32
Yellow	28

Use the information in the table to draw a bar chart in your Investigation Notebook like the one shown below. One bar has been completed for you.

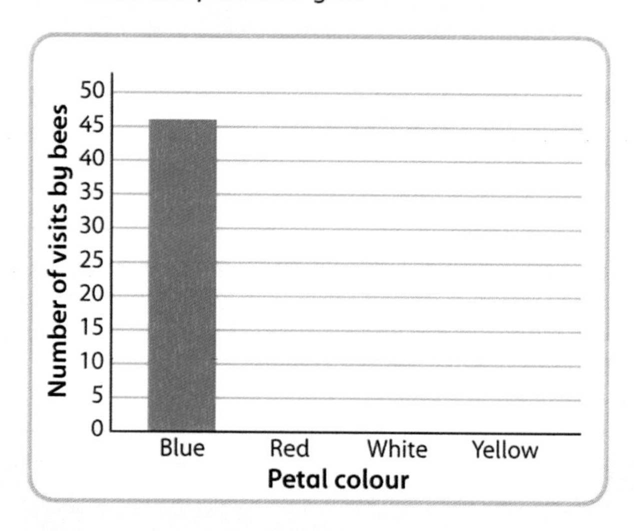

Which colour do the bees like the most?

Which colour do the bees like the least?

What do flowers provide for insects such as bees?

 What two things can be done to make sure that this is a fair test?

Part 2: Petals or nectar?

The scientists wanted to find out more about why bees are attracted to flowers. They know that bees like to drink nectar, but would nectar alone be enough to attract them? They decided to test this out.

Nectar is a sticky, sweet liquid produced by flowers

The scientists again used artificial flowers, but this time they added nectar to some and removed the petals from others. They placed the flowers outside on a sunny day and watched what happened.

Nectar and petals	Number of visits by bees
Nectar and petals	48
Petals but no nectar	35
Nectar but no petals	4
No nectar; no petals	1

Use the results in the table to create a bar chart in your Investigation Notebook.

Which do bees find more attractive – petals or nectar?

Using the information in the bar chart, explain why you think this.

If bees are attracted by petals and nectar, why did the scientists use a flower without petals or nectar?

What conclusion can we draw from this investigation?

Decide whether the following statements are true or false.

		True/False
1	Insects are the only things that can pollinate a flowering plant.	________
2	Flowers do not attract insects.	________
3	Pollen can be transported to another plant in many ways.	________

Now turn to pages 82–3 to review and reflect on what you have learned.

Looking at flowers in detail

Learn that plants produce flowers which have male and female parts, and that seeds are formed when pollen from the male part fertilises the ovum (female).

The Big Idea

Flowers have male and female parts. The plant needs these different parts to produce seeds.

1

3 Leaf

2

4

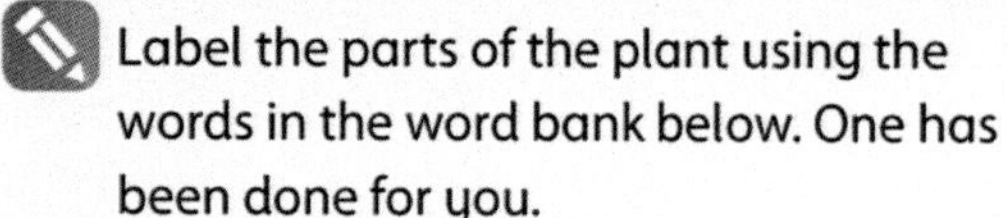

Label the parts of the plant using the words in the word bank below. One has been done for you.

Word Bank

~~Leaf~~ Stem Flower Roots

Why is each part of the plant important?

Let's look at a flower in more detail.

In a flower there are **male** and **female** parts:

The stamen

The stamen is the male part of the flower and is made up of the filament and the anther. The filament holds the anther in place and the anther is where the pollen is produced.

The carpel

The carpel is the female part of the flower and is made up of the stigma, the style and the ovary. The stigma is very sticky. This helps it to catch pollen. The style holds the stigma in place and the ovary is where the seeds are made.

Look at the detailed picture of a flower. Use the information to complete the labels for the flower using the words in the word bank below. One has been done for you and the first letters have been given.

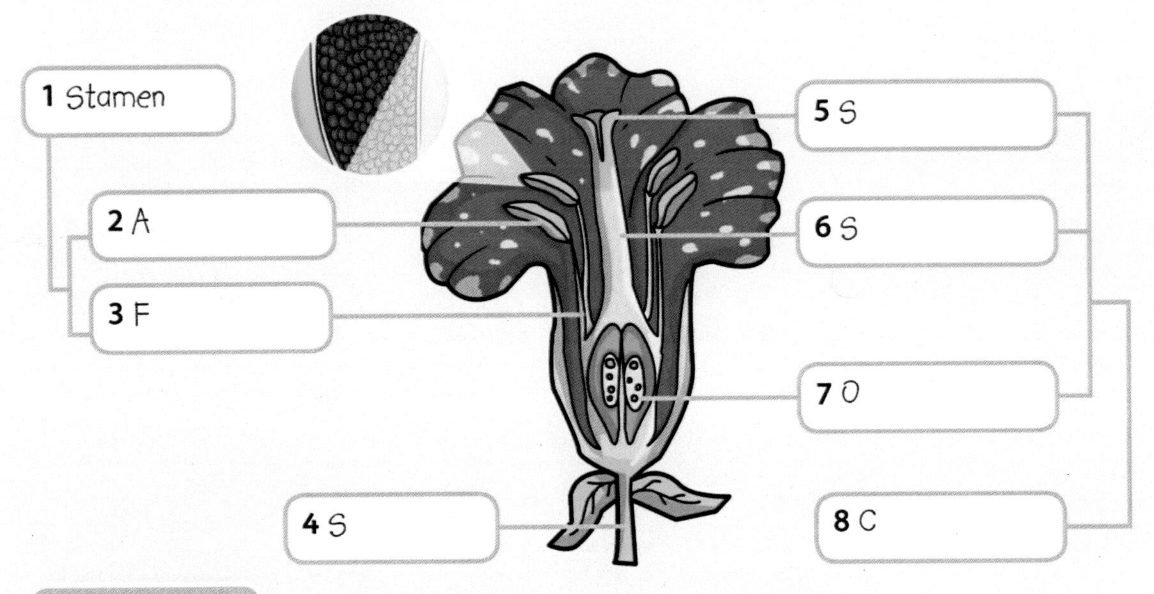

Word Bank

~~Stamen~~ Anther Style Carpel Stigma Filament Ovary Stem

Write the names of the male stamen parts and the female carpel parts in the table below. One has been done for you. Draw the stamen and the carpel in the spaces provided.

Male	Drawing of stamen	Female	Drawing of carpel
		Style	

Complete the paragraph using the words in the word bank. One has been done for you.

Flowers have male and __________ parts. The male part is called the __________, which has the anther and filament. The female part is called the carpel, which has the stigma, __________ and ovary.

Word Bank

female stamen ~~carpel~~ style

Looking at flowers in detail

Learn that plants produce flowers that have male and female parts, and that seeds are formed when pollen from the male part fertilizes the ovum (female).

The Big Idea

Pollen is used to make seeds.

What is pollination?

Pollination is when the pollen from the male anther is moved on to the sticky stigma of the female carpel. We know that there are different ways that pollen is moved.

Name one way pollen is moved from one plant to another.

Pollination can also happen through self-pollination. Look closely at the picture below. You can see that the pollen from the male anther is being dropped on to the sticky female stigma.

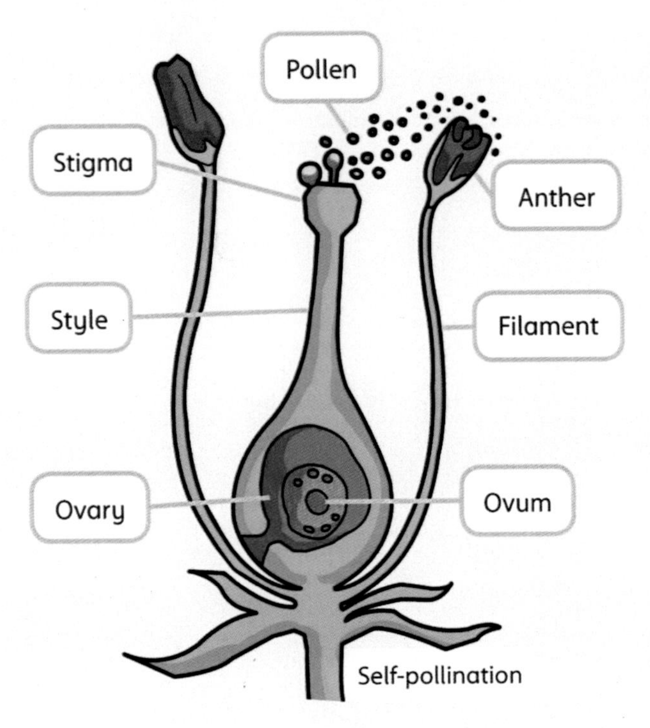

Self-pollination

The flower may be pollinated in lots of different ways but the next step is always the same. Let's look at what happens once the pollen is moved from the male anther to the female stigma.

Look closely at the picture. Can you see the pollen at the top of the stigma?

Pollen is very clever because it attaches to the sticky stigma. The pollen then sends a pollen tube down the female style. The pollen tube allows the pollen to move down the style. The pollen tube connects with the ovary and the pollen enters the ovary. When the pollen meets the **ovum** they join together and we say that **fertilisation** has happened and a seed is made.

Pollen
Stigma
Style
Pollen tube
Ovary
Ovum

Fertilisation

We can think of the pollen as going on a journey. Complete the stages of the pollen's journey using the words in the word bank below. One has been done for you.

Pollen starts its journey in the ______________. It is moved to the sticky ______________ by different methods. This is called ______________. The pollen sends a pollen ____tube____ down the ______________ and the pollen tube allows pollen to move down it. Once the pollen reaches the end of the pollen tube, it enters the ______________. When the ovum and the pollen join together ______________ happens and a ______________ is made.

Word Bank

pollination **stigma** **ovary** ~~**tube**~~ **fertilisation** **anther** **seed** **style**

Now turn to page 83 to review and reflect on what you have learned.

The big picture

Know that flowering plants have a life cycle which includes pollination, fertilisation, seed production, seed dispersal and germination.

The Big Idea

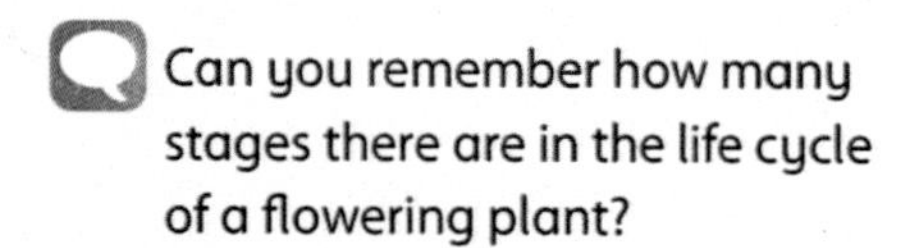

The life cycle of flowering plants has lots of big words and ideas. Let's put them all together!

Can you remember how many stages there are in the life cycle of a flowering plant?

The main stages within the life cycle of the flowering plant are seed, seedling, young plant and adult plant. Since we learned this, we have been given some new ideas to think about. We have learned that flowering plants need to be pollinated.

At what point in the life cycle is the flowering plant pollinated?

You may remember that somehow the pollen needs to reach the ovary of the female part of the flower. The pollen lands on the sticky stigma and then travels down the style until it reaches the ovary.

What is the name of the process when the pollen meets the ovum?

Once the ovum has been fertilised, a seed is produced. The seed needs to find a place to grow.

What happens to the seed?

List three ways in which a seed can be dispersed.

A seed needs to find a place that provides all the things it needs to grow. The seed needs warmth and water and then it starts the next part of its journey through the life cycle.

When the seed starts to grow we say the seed has …

Look at the picture of the life cycle. Think about the processes we have learned about: pollination germination, fertilisation, **seed production** and seed dispersal.

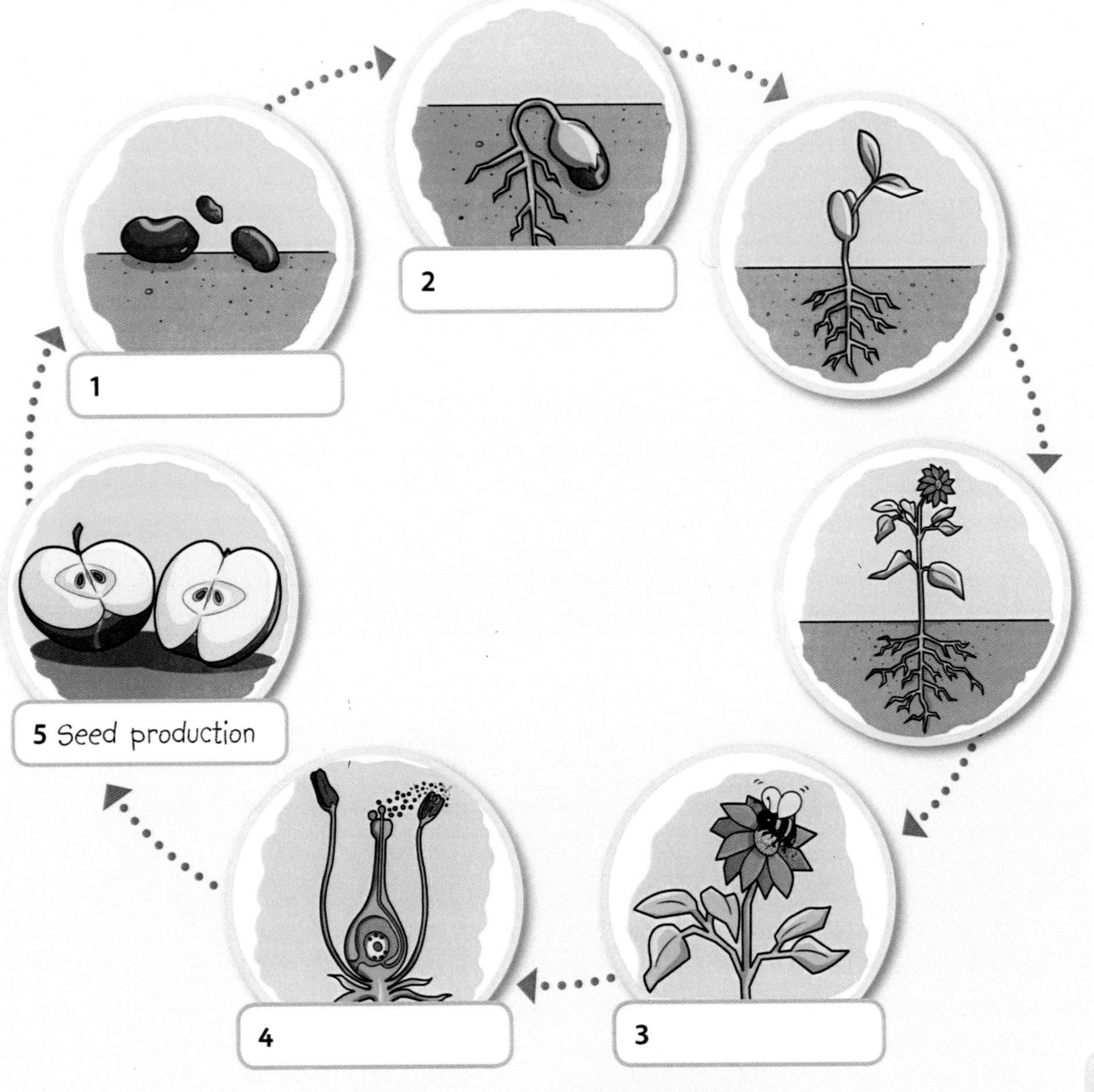

Write the name of each process in the correct place on the picture of the life cycle. Use the words in the word bank. One has been done for you.

Word Bank

Pollination Germination Fertilisation

Seed dispersal ~~Seed production~~

The big picture

Know that flowering plants have a life cycle which includes pollination, fertilisation, seed production, seed dispersal and germination.

The Big Idea

We can link everything together!

Find all ten words shown in the wordsearch.

B	O	D	N	O	I	T	A	N	I	M	R	E	G	T
P	P	H	O	S	I	E	L	C	Y	C	E	F	I	L
O	L	L	Q	D	I	P	C	T	M	O	F	R	K	Y
L	S	D	J	E	P	V	B	F	N	Z	X	D	R	A
L	K	R	E	E	O	O	Y	D	V	A	N	L	E	G
I	Y	C	E	S	P	N	E	B	C	J	L	T	P	M
N	I	W	W	W	Y	S	X	D	S	U	H	P	R	S
A	F	M	G	T	O	L	K	O	Y	V	V	X	O	Y
T	K	K	C	M	N	L	M	F	Z	U	C	W	D	T
I	V	T	W	N	O	E	F	F	U	R	M	N	U	C
O	W	C	B	S	X	Y	Q	I	J	T	U	W	C	F
N	O	I	T	A	S	I	L	I	T	R	E	F	T	Z
S	T	C	E	S	N	I	W	A	C	S	T	X	I	M
Q	L	A	S	R	E	P	S	I	D	K	B	V	O	Q
V	M	C	J	O	Z	L	P	S	J	O	J	R	N	X

 Create a life cycle of your favourite plant!

 1 How does a flowering plant produce seeds?

2 How are seeds dispersed?

3 How are flowers pollinated?

Now turn to page 83 to review and reflect on what you have learned.

What we have learned about the life cycle of a flowering plant

Flowering plants (pages 62–5)

What are the four stages of a plant's life cycle?

How does a plant's life begin?

What does a seed need in order to survive and grow?

What is the scientific word for the early growth of the seed?

I know the stages in the life cycle of a flowering plant. ◯

I know that a life cycle can be represented by a circle. ◯

Seeds, seeds everywhere! (pages 66–9)

Why are seeds dispersed?

Name four ways in which seeds are dispersed.

I understand why seeds need to be dispersed. ◯

I can describe four different ways that seeds are dispersed. ◯

Pollinating flowers (pages 70–3)

How can pollen move from one plant to another?

Insects carry pollen from one plant to another. Why do they do this?

I can describe how insects pollinate plants. ◯

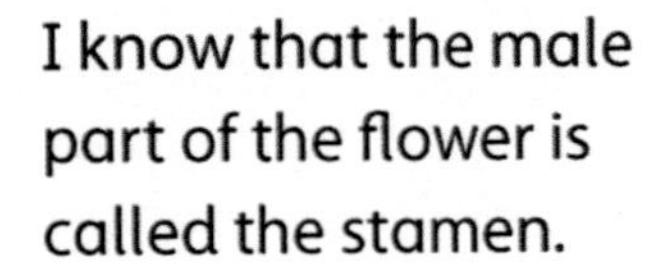

I understand that nectar attracts pollinators.

Looking at flowers in detail

(pages 74–7)

The flower contains the male and female parts of the plant. What is the scientific name for the male part? What is the scientific name for the female part?

Pollen is transferred from the male part of one plant to the female part of another plant. What is the scientific name for this process?

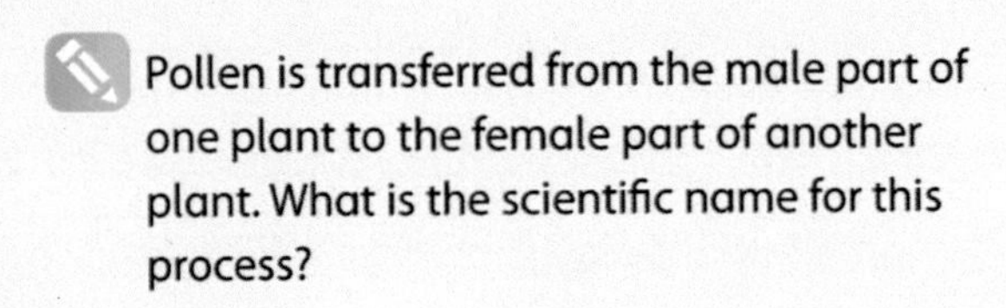

What is the scientific name for when the pollen enters the ovary?

What is produced when the ovum is fertilised?

I know that the female part of the flower is called the carpel.

I know that the male part of the flower is called the stamen.

I can describe how fertilisation happens in the flowering plant.

The big picture (pages 78–81)

Complete the paragraph using the words in the word bank below.

The life of a flowering plant begins with __________________ dispersal. It then __________________ and grows into an adult plant. Later, it is then pollinated by ______________. After __________________ a new seed is produced.

Word Bank

insects	fertilisation
seed	germinates

I understand why the life of a flowering plant is called a life cycle.

4 Investigating Plant Growth

In this module you will:

- find out about the conditions seeds need to germinate
- find out about the conditions plants need to grow well
- learn how to investigate like scientists.

Word Cloud

light energy

energy

warmth

Amazing fact

The largest trees in the world are the Giant Redwoods. They grow to a height of over 100 metres and have a circumference of 20 metres!

Seeds grow upwards from under the ground and continue growing upwards towards the sun.

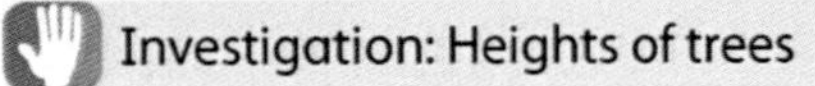

Investigation: Heights of trees

1 Look around where you live and guess how high the trees are. Are the trees growing on their own or are they close to other trees? If you do not have any trees close by, look at local plants growing.

2 Compare the heights of the trees/plants growing on their own with those growing close to each other. Try to compare trees/plants of the same type (species). If the leaves are the same, then the tree/plant will be of the same type.

Copy and complete the table below in your Investigation Notebook.

Height of trees/ plants growing on their own	Height of trees/plants growing close to other trees/plants

Why are the trees growing together taller than those which are growing on their own?

Investigating seed germination

Find out that seeds need water and warmth to germinate, but not light.

The Big Idea

For life to begin, seeds need the right conditions.

After seeds have been dispersed they begin the next stage of the life cycle, which is germination. This is where the seed develops and grows under the ground and then breaks through the surface into the air to become a seedling.

Sunny garden

Well-watered lawn

Concrete path

Desert

Oasis

 Look at the pictures above. Choose two places where plants are growing. Why do they like growing there?

 Choose two places where plants are not growing. Why do they not like growing there?

In the pictures above there is plenty of water but there are no plants growing. Why?

Think about where you live. Do new plants appear at certain times of the year? If so, during which season does the new plant growth start? Why?

Scientists first look at situations, such as the places shown in the pictures on the left, and try to understand what they tell us. You have already done this and using your observations you have predicted that:

1 plants only grow well when they have fertile soil and water

2 plants do not grow well when they have fertile soil but are in cold conditions

3 plants start to grow in spring when it is warmer.

Using the evidence in points 1–3 above, scientists can now predict that plants will only grow well if they have fertile soil, water and **warmth**.

To be sure, scientists have to test their predictions to see if they are correct. In order to test their predictions they have to plan an investigation.

Scientists investigate using scientific enquiry methods that are based on observation and prediction. We can describe this enquiry process in simple steps:

- **Observation:** look and notice differences and similarities.
- **Prediction:** from your observations try to explain what you have observed.
- **Question:** ask yourself questions like 'I wonder why …?' or 'I wonder what would happen if …?'
- **Investigation:** investigate the questions you have raised using scientific enquiry methods.

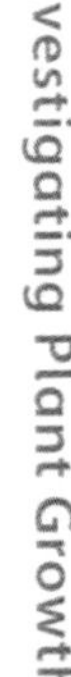

Investigating seed germination

Find out that seeds need water and warmth to germinate, but not light.

The Big Idea

Scientists can investigate which are the right conditions for seeds to germinate.

What are the conditions in which seeds will germinate best?

Using the information in the previous unit, a scientist would predict that seeds planted in fertile soil will not germinate very well:

- without water
- in the cold.

Investigation: How important are water and warmth to seed germination?

In your investigation, you will need four pots, each containing ten grass seeds planted in the same amount of fertile soil.

Why is it important to put the same number of seeds in each pot?

Below is a list of all the variable combinations you will need to test:

- Pot 1: Both water and warmth.
- Pot 2: Water but no warmth.
- Pot 3: No water but warmth.
- Pot 4: No water and no warmth.

To make sure that the seeds do not have warmth, we can put them in a fridge.

Circle the correct words in each statement below to give you the correct test conditions for pots 1–4. The first one has been done for you.

Place Pot 1 in the classroom/fridge with water/no water.

Place Pot 2 in the classroom/fridge with water/no water.

Place Pot 3 in the classroom/fridge with water/no water.

Place Pot 4 in the classroom/fridge with water/no water.

1. Place the same amount of soil in each of your four pots and plant ten grass seeds in each.
2. Label and date each pot.
3. Place each pot in one of the four different conditions. Do not overwater the seeds that require water. (Sprinkle with water every three or four days so they do not dry out.)
4. Observe the pots weekly.

In your Investigation Notebook, copy and complete the table below by writing 'Grown' or 'Not grown'. When the grass appears above the surface of the soil, you can write 'Grown'.

	Grown/Not grown			
	Pot 1	Pot 2	Pot 3	Pot 4
Week 1				
Week 2				

Measure your seedlings each week. In your Investigation Notebook, copy and complete the table below.

	Height (cm)			
	Pot 1	Pot 2	Pot 3	Pot 4
Week 1				
Week 2				

Which pot contains the tallest grass? Why?

Complete the paragraph below.

From my investigation, I have learned that seeds need ________________ and ________________ in order to germinate successfully.

1 What is the scientific name for when a seed starts to grow?

2 What is the best time of year for plants to germinate?

3 When scientists investigate they arrange the conditions so they can see what changes happen. This is known as ________________ the variables.

Now turn to pages 98–9 to review and reflect on what you have learned.

Let there be light!

Understand that plants need energy from light to grow.

The Big Idea

Plants are made up of different parts and each one is important in their growth and reproduction.

Look at the plants growing outside or those in the picture above. Can you remember the names of the different parts of the plant?

 Label the diagram using words from the word bank below. One has been done for you.

Word Bank

Leaf ~~Flower~~ Stem Root

Look carefully at the picture of a plant or a real plant. Scientists do more than look, they observe. This means that they notice detail and ask questions such as 'I wonder why …?'

 Ask yourself the questions below. Write your suggestions in the spaces provided. Try to use some of the words in the word bank.

Word Bank

reproduce	**pollination**	**food**
transport	**fertilisation**	

I wonder why a plant has roots?

I wonder why a plant has a stem?

I wonder why a plant has leaves?

I wonder why a plant has a flower?

 Investigation: How can we find out what plants need to grow well?

Look at the pictures provided by your teacher or go on a nature walk.

1 Notice how well the plants grow in the Sun compared with those in the shady areas.

2 Notice how well the plants grow in the damp soil compared with those in dry areas.

 What predictions can you make from your observations?

I think that plants grow better when they have ________________.

I think that plants grow better when they have ________________.

Let there be light!

Understand that plants need energy from light to grow.

The Big Idea

Water and light are important to the growth of green plants.

Investigation: The importance of light and water to plant growth

In the previous unit we learned that plants do not grow well in the shade or in very dry conditions. We predicted that light and water are very important to plant growth. As scientists, we can set up an investigation to compare and measure how important these two variables are to growth.

Scientists try to observe 'cause and effect'. In this case, we look at what causes the growth (light and water) and what the effect is if the causes are removed.

In our previous investigation of how important water and warmth are to seed germination, we worked out how to control the variables so we could see the effect of each one. We can do the same for this investigation, but with water and sunlight as the variables.

1 Your teacher will give you four pots containing grass. Here is a list of all the variable combinations you will need to test:

- Grass pot 1: Both sunlight and water.
- Grass pot 2: Sunlight but no water.
- Grass pot 3: No sunlight but water.
- Grass pot 4: No sunlight and no water.

Complete the table by writing 'Yes' or 'No' under the headings 'Sunlight' and 'Water' to show the test conditions each grass pot will be given to test all four variable combinations. The first one has been done for you.

	Sunlight	Water
Grass pot 1	Yes	Yes
Grass pot 2		
Grass pot 3		
Grass pot 4		

2 Test the grass samples in the different conditions over the next two weeks.

The condition of water or no water is easy to arrange. How can we arrange the condition of light or no light?

When scientists compare and test different conditions to see which is best, they test one of the samples under normal conditions. This tells them how good or how bad the changes are compared with normal. Scientists call this sample kept under normal conditions a control.

Which one of your samples is the control?

Look at the samples weekly and measure how much they have grown. In your Investigation Notebook, record your results in a table like the one below.

Date	Height (cm)			
	Grass pot 1	Grass pot 2	Grass pot 3	Grass pot 4

What do your findings show?

In this investigation we discover that the plants left in darkness die fairly quickly. Our conclusion as scientists is that light is essential for the growth of plants.

Why is light so important to plants?

Would you die if you were kept in the dark for two weeks?

All living things need **energy** to live. Animals get their energy from the food they eat.

Where do plants get their energy from?

Let there be light!

Understand that plants need energy from light to grow.

The Big Idea

Light is very important for plant growth.

Investigation: Do plants grow better in more sunlight? (Part 1)

We have already observed that plants grow better in sunlight and we correctly predicted that sunlight is essential to plant growth. Our curiosity as scientists leads us to another question:

- Do plants grow better if they have more sunlight?

We are going to think like a scientist and plan an investigation using scientific enquiry methods.

The two variables that we need to consider in this investigation are:

- the amount of light
- the amount of growth.

These are the two most important things we need to measure. How much *light* produces how much *growth*?

However, this now introduces a third variable:

- time.

We need to measure the change of growth over *time*.

Scientists have to make sure that the test is fair. This means that all the samples must have an equal amount of time to grow. The tests must begin at the same time and end at the same time.

It is also important to make the test fair by making all the test samples equal. This means that the seeds or plants we are growing all have to be the same kind and they need to grow in the same place and in the same type and amount of soil. They must also be watered at the same time using the same amount of water.

In this investigation we are going to grow grass. When the grass begins to surface we can begin our investigation. The grass seeds are starting with zero growth so the investigation is equal and fair.

1 Set up four samples, with Sample 1 having no light and Sample 4 having full sunlight. Sample 4 will be your control.

Decide how much sunlight to give Samples 2 and 3 and complete the table below. Samples 1 and 4 have been done for you.

Number of sunlight hours	Sample 1	Sample 2	Sample 3	Sample 4
	0			12

Colour in the fraction of sunlight for Samples 1 to 4 in yellow in the circles below. One has been done for you.

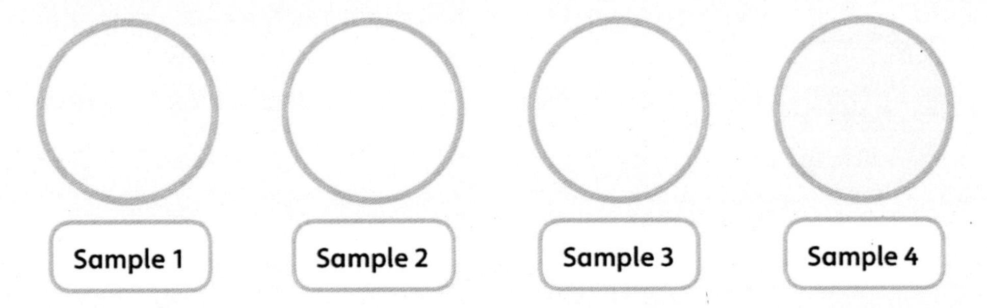

2 Decide how long you will allow your samples to grow. Remember to water them carefully every two to three days.

Measure the height of your grass samples every day. In your Investigation Notebook, record your results in a table like the one below.

Day	Height (cm)			
	Sample 1	Sample 2	Sample 3	Sample 4
1				
2				
Total growth (cm)				

Let there be light!

Understand that plants need energy from light to grow.

The Big Idea

Numerical results are easier to understand if we can change them into pictures.

 Investigation: Do plants grow better in more sunlight? (Part 2)

In the previous unit we measured and recorded the growth of each sample daily. In this second part of the investigation we are going to use those results to make a graph.

 In your Investigation Notebook, draw outlines for four graphs like the one shown below. Plot a different coloured line on each graph for one of your four samples.

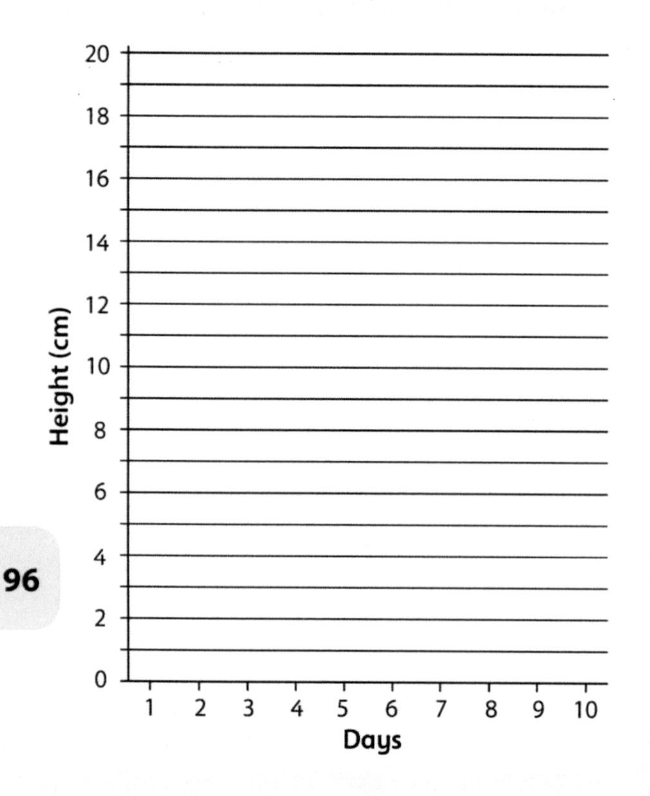

Scientists look at the results and graphs of an investigation and interpret what they mean.

 What do the results in your line graphs show?

Copy and complete the write-up of the investigation below.

Do plants grow better in more sunlight?

What we did:

We decided to measure the growth of grass over ________ days.

We decided to test four samples of grass in four different conditions of sunlight.

- Sample 1 was placed in the dark so it had no sunlight.
- Sample 2 was placed in the sunlight for ________ hours each day.
- Sample 3 was placed in the sunlight for ________ hours each day.
- Sample 4 was placed in the sunlight for 12 hours each day.

We measured the height of the grass using string and a centimetre ruler. We did this for each sample every day at the same time for ________ days.

We recorded the measurements in a results table.

We then plotted line graphs of the growth of each sample of grass over the time of the investigation.

What we found out:

The longest grass was the sample with ________ hours of sunlight each day. It was ________ centimetres long.

The next longest grass was the sample with ________ hours of sunlight each day. It was ________ centimetres long.

The next longest grass was the sample with ________ hours of sunlight each day. It was ________ centimetres long.

The shortest grass was the sample with 0 hours of sunlight each day. It was ________ centimetres long.

Conclusion:

This investigation proves that ...

Think about...

Green plants are the only living things that can change the Sun's energy into food. The only source of energy for the Earth is the Sun. What will happen if we do not look after our plants on Earth?

1 What are the two most important things for plant growth?

2 In some countries, and at some times of the year, plants do not grow well even though they seem to have all they need. Why?

3 What is it in sunlight that gives life to plants?

Now turn to page 99 to review and reflect on what you have learned.

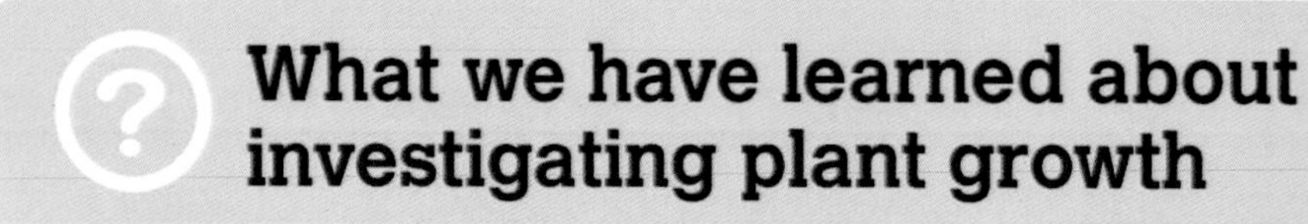

What we have learned about investigating plant growth

Investigating seed germination

(pages 86–9)

How do we know that seeds do not need light to germinate?

The four stages of a scientific investigation are:

- prediction
- question
- investigation
- observation.

Put these stages in order from the beginning to the end of the scientific enquiry process.

I understand why plants only need water and warmth to germinate. ◯

I know the four stages of the scientific enquiry process. ◯

Let there be light! (pages 90–7)

If we are investigating the effect of light on plant growth, what variables do we need to keep the same in order to make the test fair?

In this unit we learned how to record our findings in the form of a graph. Graphs make it easier for us to interpret our findings and draw conclusions. What kind of graph shows us how things change over time?

I know that green plants use **light energy** to make them grow. ◯

I understand how to control variables. ◯

5 Earth's Movements

In this module you will:

- explore that the Sun does not move; it appears to move because of how the Earth spins on its axis
- know that the Earth spins on its axis once every 24 hours
- know that the Earth takes a year to orbit the Sun, spinning as it goes
- research the life and discoveries of scientists who explored the solar system and stars.

Some people think that the Sun moves across the sky every day.

Do you think the Sun is moving and the Earth is standing still?

THE SUN IS FALLING OUT OF THE SKY LATER TODAY!

Have you ever looked up at the sky at night? We can see stars and planets. Some of the stars are so far away that it takes time for the light to reach our eyes. We see the Moon because it is reflecting light from the Sun. When we see the Moon we are seeing it as it was one second ago. When we see the Sun we are seeing it as it was about eight seconds ago. Some planets are so far away that if people on the planet were looking at us now they would see the pyramids being built!

Why can we see the Moon?

Try spinning around. How long does it take you to feel dizzy?

The Earth spins at over 1600 kilometres an hour.

Why do we feel dizzy when we spin, but not when the planet we are on spins?

Our solar system is very big. It would take over 30 **years** to travel to the outer edge of the solar system on the fastest rocket. The planets travel around the Sun. It takes one year for the planets to complete one circle around the Sun. Each planet has a year of a different length.

Why is a **Mercury** year shorter than an Earth year?

Why is a **Jupiter** year longer than an Earth year?

The Sun appears to move, but it doesn't

Explore through modelling that the Sun does not move. Its apparent movement is caused by the Earth spinning on its axis.

The Big Idea

The Earth is spinning and moving through space.

 List five **planets** in our solar system.

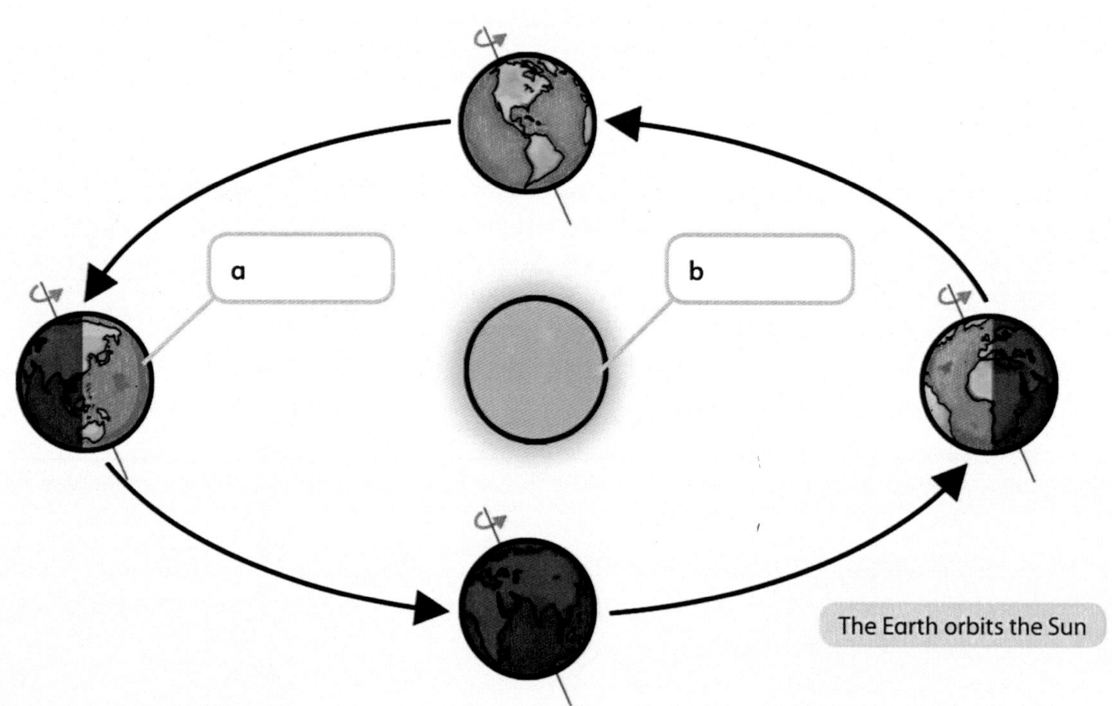

The Earth orbits the Sun

 Label the diagram above using the words in the word bank below.

Word Bank

Sun Earth

The Sun is a **star**. A star produces its own light. Planets do not produce their own light. We can see them because they reflect the Sun's light.

The Sun is in the middle of the solar system. The **Earth** is the third planet away from the Sun. The heat and light the Sun gives the Earth make it perfect for life.

The Moon is much smaller than the Sun and the Earth. We could fit almost four Moons into the Earth. The Moon is not a planet. It is too small and is a satellite of the Earth. A satellite is an object that orbits around another object.

 What is the difference between a star and a planet?

 Is the Moon a planet or a satellite?

 How does the Earth's distance from the Sun allow life to survive?

It is very difficult to imagine the sizes of stars and planets because they are so big. Sometimes **scientists** model things that they cannot see clearly:

- The Moon could be shown as a bead.
- The Earth could be shown as a tennis ball.
- The Sun could be shown as a football.

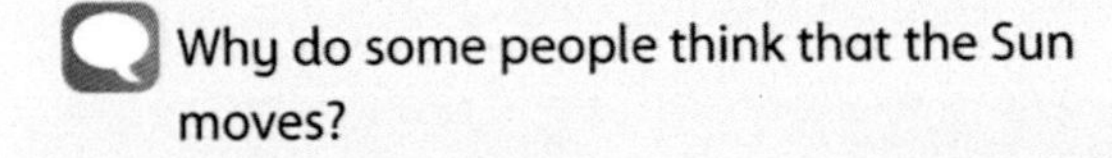
Why do some people think that the Sun moves?

You might have heard people say that the Sun rises in the east and sets in the west. This makes you think that the Sun is moving. The Sun does not move. It is really the Earth spinning on its own **axis**.

An axis is an imaginary line through the Earth. It is like pushing a stick through your tennis ball. At one end of the axis there is the North Pole and at the other end the South Pole.

The Earth spins on its axis but that is not the only movement it makes.

The Earth also moves around the Sun. The path of the Earth around the Sun is called its **orbit**. All of the planets in our solar system orbit the Sun. The Moon orbits the Earth.

 What is an orbit?

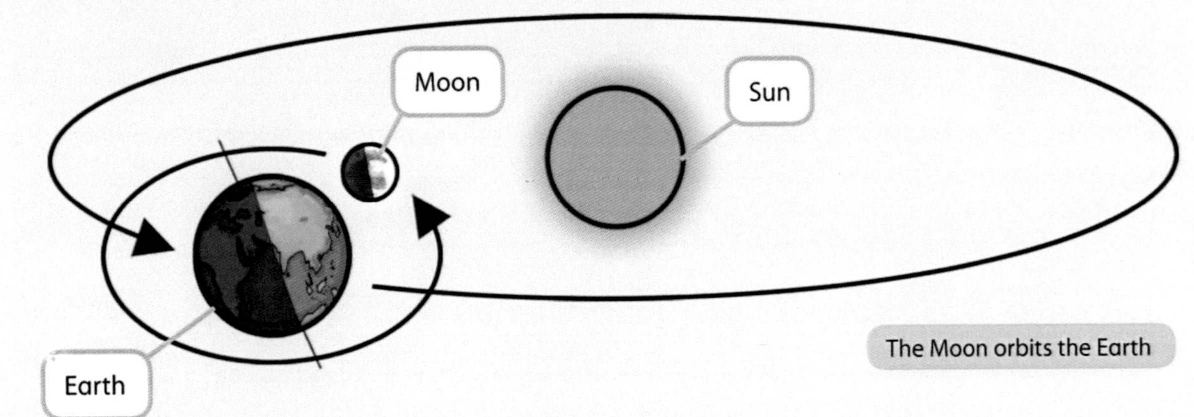

The Moon orbits the Earth

The Sun appears to move, but it doesn't

Explore through modelling that the Sun does not move. Its apparent movement is caused by the Earth spinning on its axis.

The Big Idea

The Sun seems to move, but it doesn't.

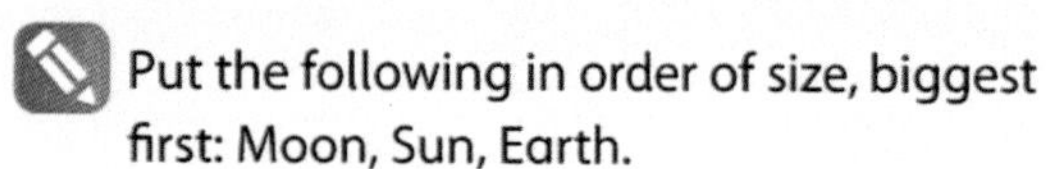

Put the following in order of size, biggest first: Moon, Sun, Earth.

Which ones move in the solar system?

The Sun seems to move from one side of the Earth to the other.

The apparent movement of the Sun can be captured on film and in photographs. It is hard to believe that the Sun isn't moving in the solar system. It is fixed in the middle. It is actually the Earth spinning that makes it look like the Sun is moving. The Earth makes a complete turn every **24 hours**.

One half of the Earth is always in shadow and the other half is lit by the Sun. We have daylight when the place where we live is turned towards the Sun. We have night-time when the place where we live is turned away from the Sun.

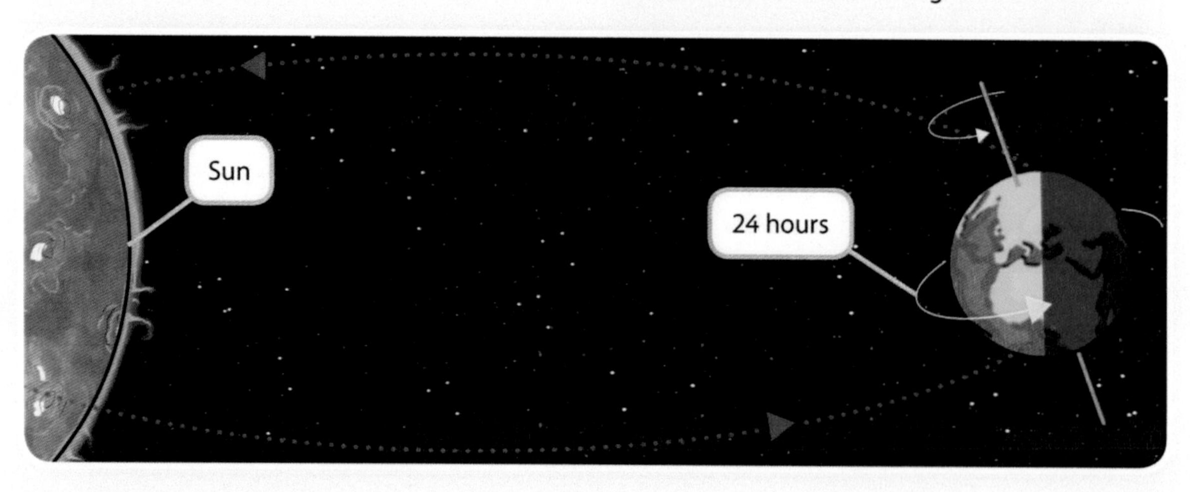

Investigation: Where is the Sun shining in your classroom at the moment?

1 Your teacher will give you a simple diagram of your classroom clearly showing the windows.

2 Mark on your diagram which window the Sun is shining through. Write the time of day on the diagram.

3 Keep a record of where the Sun is shining every 30 minutes for the rest of the school day. Number the windows and record your results in your Investigation Notebook in a table like the one shown below.

4 Record your observations of the Sun. Think about the following questions to help you:

- Can you feel the heat from the Sun?
- How far does the light shine into the room?

Window	Time of day	Observations
1		
2		
3		

Look at your diagram and table of results. What do you notice about the sunlight?

Does the light come through different windows at different times of the day?

Can you use your table of results to predict the time?

You could test this prediction on the following days after the investigation. Look at the window the Sun is shining through. Now look at your results. What time of day did it shine through that window in your investigation? Check to see if it is the same time.

Is this a reliable way to tell the time?

Explain why your results might not be reliable.

What can you do to make sure you collect reliable results?

Sometimes, scientists compare their results with other people's. This is to make sure their results are very reliable. The results they collect from other people are called secondary data. Remember, the more results you collect that fit the same pattern, the more reliable the results are.

How can you use secondary data here?

Repeat this investigation over the next few weeks.

The Sun appears to move, but it doesn't

Explore through modelling that the Sun does not move. Its apparent movement is caused by the Earth spinning on its axis.

The Big Idea

The Sun does not move.

To stay cool in the Sun, we can sit under a tree or a sunshade. If we stay there all day, we will have to move to stay out of the Sun. Why?

The Sun does not move across the sky. It is the Earth that is spinning on its axis. This gives the impression that the Sun is moving.

If a 10-year-old child travels to the Sun by rocket, they will be 19 years old when they get back. So we will observe the Sun from here on Earth!

Investigation: Measuring the length and position of a shadow to show the apparent movement of the Sun

You are going to carry out an investigation on the Sun using shadows.

1 Find a flat, clean and safe space outside.

2 Sit on the floor and observe the length and width of your shadow.

3 Stand up and ask your partner to draw around your shadow.

4 Swap roles so that your partner stands up and you draw around their shadow.

5 Measure the length and width of your shadows.

6 Take it in turns to lie down next to your shadow. Ask your partner to compare you with your shadow. Record your results in your Investigation Notebook in a table like the one shown opposite.

Is your shadow smaller or bigger than you? Is it wider or thinner than you? Or is it exactly the same size as you?

To observe the apparent movement of the Sun, scientists sometimes measure how high the Sun is in the sky.

7 Sit on the ground with the Sun in front of you.

⚠ Never look directly at the Sun. It will damage your eyesight.

8 Hold up a ruler so that the bottom is level with the horizon.

9 Looking out of the corner of your eye, estimate the height of the Sun using the ruler.

10 Record your results in the same table in your Investigation Notebook.

This girl is looking out of the corner of her eye

You could repeat this investigation at different times during the day.

Time of day	Length of shadow (cm)	Width of shadow (cm)	Height of the Sun (cm)	Observations

Does the length of your shadow depend on the time of day?

How does the shape of your shadow change?

Does the height of the Sun change throughout the day?

Is there a link between the size of the shadow and the height of the Sun from the horizon?

Circle the correct words. One has been done for you.

At midday, my shadow was longer/shorter in length than in the morning. At the end of the day, my shadow was longer/shorter than at midday. The Sun was highest in the sky at dawn/midday. It was lowest in the sky at midday/dusk.

The Sun appears to move, but it doesn't

Explore through modelling that the Sun does not move. Its apparent movement is caused by the Earth spinning on its axis.

The Big Idea

The Sun appears to rise and set every day.

The Sun rises in the morning to mark the end of night and the beginning of day. Day and night happen at different times across the world. Even where you live, the Sun rises and sets at different times during the year. Look at the information you collected in other investigations about the shape of your shadow.

On 1 March 2013, the sunrise time in Jeddah, Saudi Arabia was at 06.44. The sunset time was 18.28. The total amount of sunlight on that day was 11 hours, 43 minutes and 28 seconds.

On 31 March 2013, the sunrise time was 06.17. The sunset time was 18.38. The total amount of sunlight was 12 hours, 20 minutes and 32 seconds.

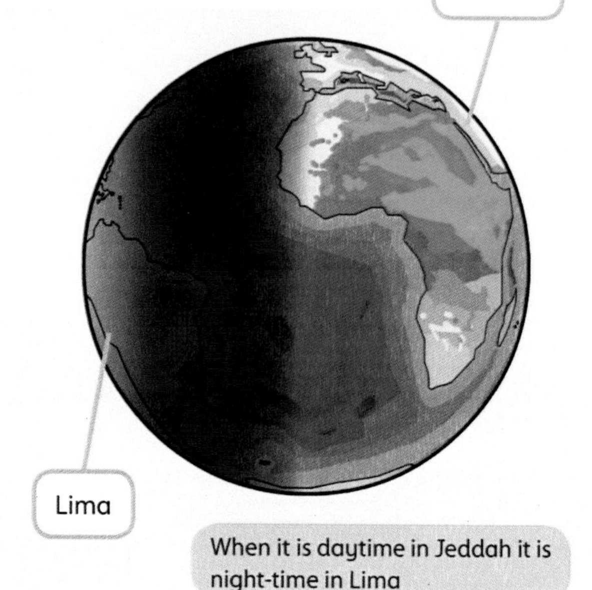

When it is daytime in Jeddah it is night-time in Lima

 Which day had the most sunlight?

Month	Sunrise time	Sunset time	Total hours of sunlight
January	7.01	17.53	10h 51m
February	7.01	18.13	11h 12m
March	6.44	18.28	11h 43m
April	6.16	18.38	12h 22m
May	5.52	18.49	12h 21m
June	5.40	19.02	13h 21m
July	5.44	19.10	13h 25m
August	5.56	19.02	13h 06m
September	6.07	18.47	12h 32m
October	6.15	18.11	11h 56m
November	6.27	17.47	11h 20m
December	6.45	17.40	10h 55m

Average sunrise and sunset times for Jeddah in Saudi Arabia in 2013

 Describe the pattern in the data.

 Do some months have more sunlight hours than others?

It is difficult to read big tables of results like this. It would be easier to see them on a graph.

 What kind of graph could you draw?

Draw a graph in your Investigation Notebook to display each month's total hours of sunlight.

 What is the independent variable?

 What is the dependent variable?

 Which month of the year has the most sunlight hours?

Which month has the earliest sunrise?

 Which month has the latest sunset?

 Are the days with the most sunlight hours at the same time of year?

 Which month has the fewest sunlight hours?

Think about...

Do sunset and sunrise data vary around the world? Do other countries witness a similar pattern in data? Are all sunlight hours in a day the same length around the world?

 1 Circle the biggest:

Sun Earth Moon

2 Describe how the Earth moves. Include the word 'axis'.

3 Explain why some people think that the Sun moves.

Now turn to page 118 to review and reflect on what you have learned.

How long does it take the Earth to spin on its axis?

Know that the Earth spins on its own axis once in every 24 hours.

The Big Idea

What makes a day a day?

How many hours are there in a day?

The Earth spins at speeds above 1600 kilometres an hour. We do not feel this because everything else on the Earth is spinning at the same speed. Although the Earth is spinning at such a high speed, it takes 24 hours for it to make a complete spin. The time it takes the Earth to make one full spin is how we measure a day.

Think about all of the things you do in a day. Remember night-time as well.

How do you know that it is morning? What sounds do you hear? What can you smell? How much sunlight is there? Then, how do you know that it is the end of the day?

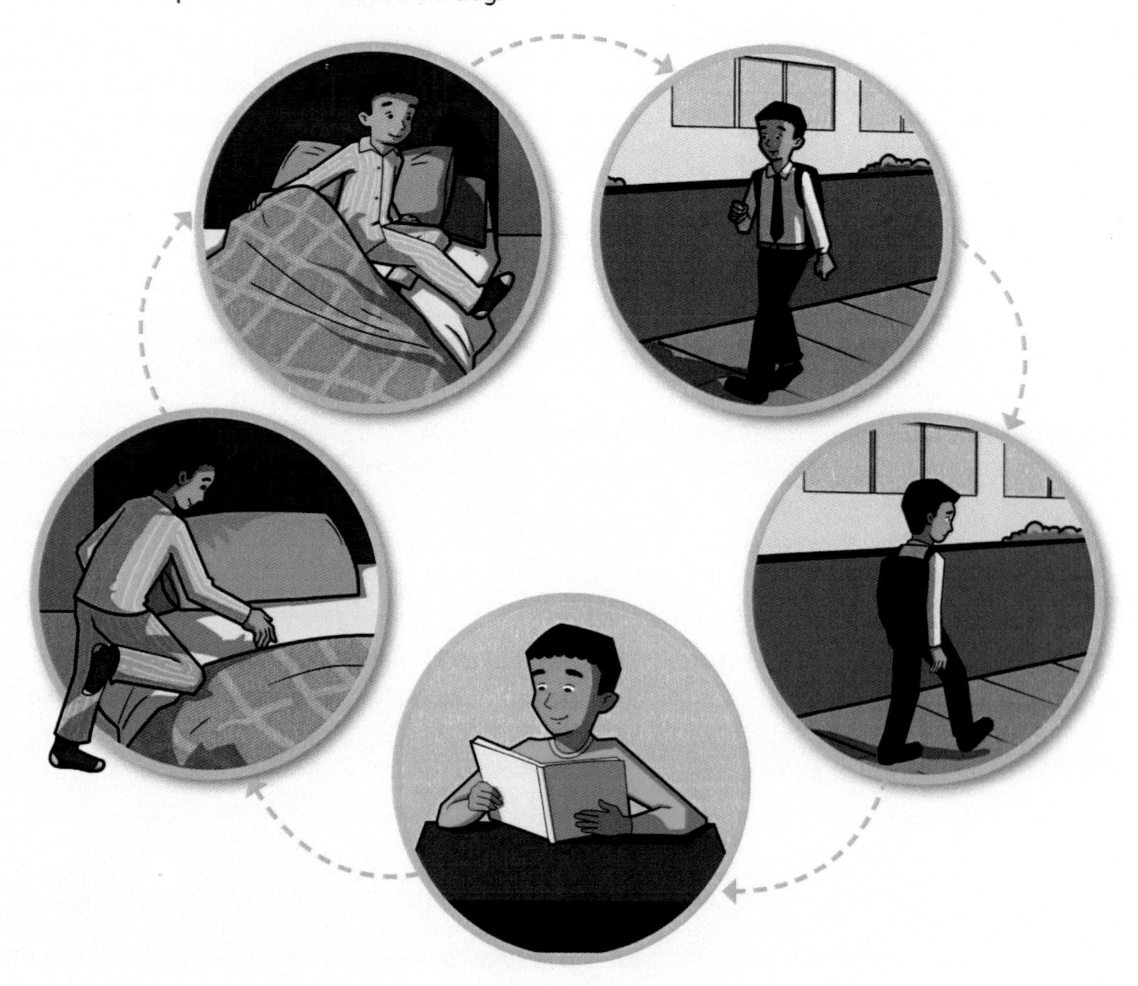

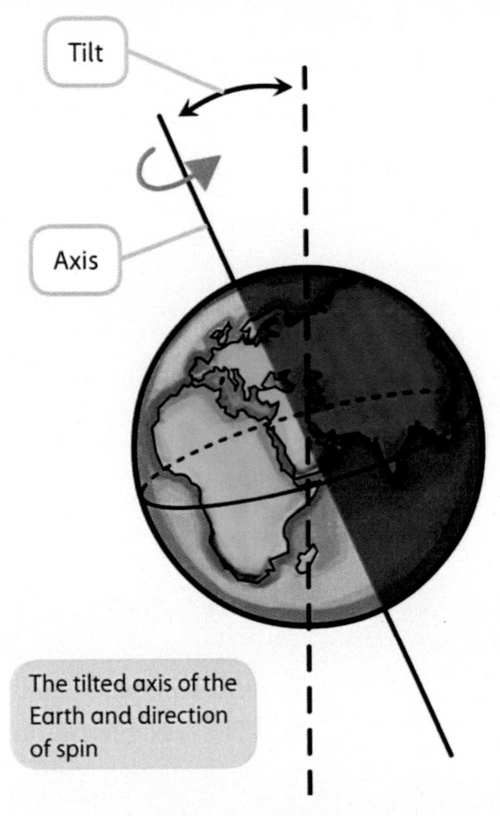

The tilted axis of the Earth and direction of spin

Look at the diagram above. Notice that the axis is slightly tilted. The blue arrow shows the direction of the Earth's spin. Certain parts of the Earth are in darkness for some of the time. This is because that part of the Earth is no longer facing the Sun.

Investigation: Modelling the Earth's spin

1 Make a model of the Earth using modelling clay.
2 Make an axis with a stick or a pencil.

 Take care when pushing the stick through the centre of your modelling clay.

3 Ask your partner to hold a lit torch in position. The torch is a model of the Sun.
4 Put a tiny ball of clay on the surface of your Earth. This is you. Hold your Earth in front of the Sun and carefully spin it.

 What happens to the sunlight on you?

 Are you in sunlight for a whole spin?

Think about...

Do all countries get the same amount of sunlight? Why are some countries hotter than others?

 1 How long does it take the Earth to spin on its own axis?

2 Explain why we have night and day.

Now turn to page 118 to review and reflect on what you have learned.

How does the Earth orbit the Sun?

Know the Earth takes a year to orbit the Sun, spinning as it goes.

The Big Idea

The Earth orbits the Sun as well as spinning on its axis.

Does the tilted axis of the Earth affect us?

How long is a year?

Can you remember what happened on your last birthday? What about the one before that? Think back to about a week or even a month ago. How much longer is a year? Look outside. How has the view changed over the past year? Have any of the buildings changed or been painted? How have any trees or plants changed?

Look at the photographs above. What changes might happen over the year?

How long does it take the Earth to spin on its axis?

While the Earth is spinning on its own axis, it is also orbiting the Sun. The Earth is held in orbit by the Sun's gravity. The Earth continuously falls around the Sun, spinning as it goes. It takes $365\frac{1}{4}$ days to do this. $365\frac{1}{4}$ days is one full year.

The tilted axis of the Earth and direction of spin

Other planets in our solar system do not have $365\frac{1}{4}$ days in their year.

It takes **Neptune** 165 Earth years to orbit the Sun. That is a long time between birthdays!

Planet	Length of year
Venus	225 Earth days
Mars	687 Earth days
Jupiter	12 Earth years
Saturn	29 Earth years
Uranus	84 Earth years
Neptune	165 Earth years

What happens to the quarter of a day?

Every four years there is a leap year. We add up all the odd quarters of a day and make them into an extra day in February. Look at the diagram above. When the Earth is facing the Sun some parts are closer to the Sun than others. This gives us the seasons of the year.

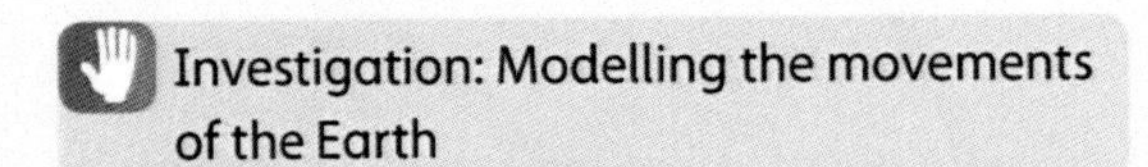

Investigation: Modelling the movements of the Earth

1 Working in pairs, ask your partner to stand very still in a space holding a lit torch. They are the Sun.

2 Stand opposite your partner and practise spinning on your axis in an anticlockwise direction. You are the Earth.

3 Now try to orbit the Sun in an anticlockwise direction at the same time as spinning anticlockwise on your axis. You are modelling how the Earth gives us a day as you spin and a year as you orbit.

4 Take turns to be the Sun and the Earth.

Think about...

How would you measure your age if you lived on Neptune?

1 How long does it take the Earth to orbit the Sun?

2 What happens to the quarter of a day?

Now turn to pages 118–19 to review and reflect on what you have learned.

The discovery of our solar system

Research the life and discoveries of scientists who explored the solar system and stars.

The Big Idea

Some people think that the Earth is flat.

How do we know what shape the Earth is? What evidence do we have?

What shape is the Earth?

How do we know what shape the Earth really is?

People have always been interested in looking at the stars and planets. There are records from just over 6000 years ago about the stars. The people who made them were called cosmologists. They studied the movements of the planets and stars. They made models of the positions of the stars and planets. They knew lots of facts about the planets, but they thought that the Earth was flat. Scientists and educated people believed this. They did not have any evidence to question it.

Aristotle was born about 2500 years ago. He had a very interesting job. He was a thinker! He started to think that everyone else was wrong about the Earth. He thought that the world was not flat but spherical. He worked very hard to get other people to believe this. It eventually became accepted that the Earth was spherical and not flat.

Aristotle

In the 19th century, an English inventor called Samuel Rowbotham decided that the Earth was flat after all. He set up a society, or group, to try to convince everyone that the Earth was flat.

How do scientists prove they are right?

Scientists have a question that they want to answer. They investigate this question and collect data to prove the answer.

How do scientists know that they can rely on the data?

Samuel repeated his tests but he did not know that his experiments were wrong, so his results were also wrong. Many other people carried on his work. They tried to convince everyone that the Earth was flat. In 1956, Samuel Shenton set up the Flat Earth Society based on Samuel Rowbotham's work. The Flat Earth Society is still active today.

Do you think it would be difficult to convince people that the Earth is flat?

Is there any evidence to prove that the Earth is a sphere?

'The world is now a very small place.' This is a common phrase.

What exactly does this phrase mean? Has the Earth shrunk?

People travel the world. They go on holiday or travel for business. If the Earth was flat, what would happen when we came to the edge?

With all this evidence it is hard to believe that the Flat Earth Society is still supported today.

Plan a leaflet in your Investigation Notebook to show younger children in your school that the Earth is a sphere. Think about the following questions to help you:

- What evidence can you use?
- Is the evidence reliable?
- Where will you get the information from?
- Can you find any evidence to prove that the Earth is flat?

A 'flat' Earth – to see a spherical Earth look at page 117

The discovery of our solar system

Research the life and discoveries of scientists who explored the solar system and stars.

The Big Idea

Space technology has improved over the last 50 years.

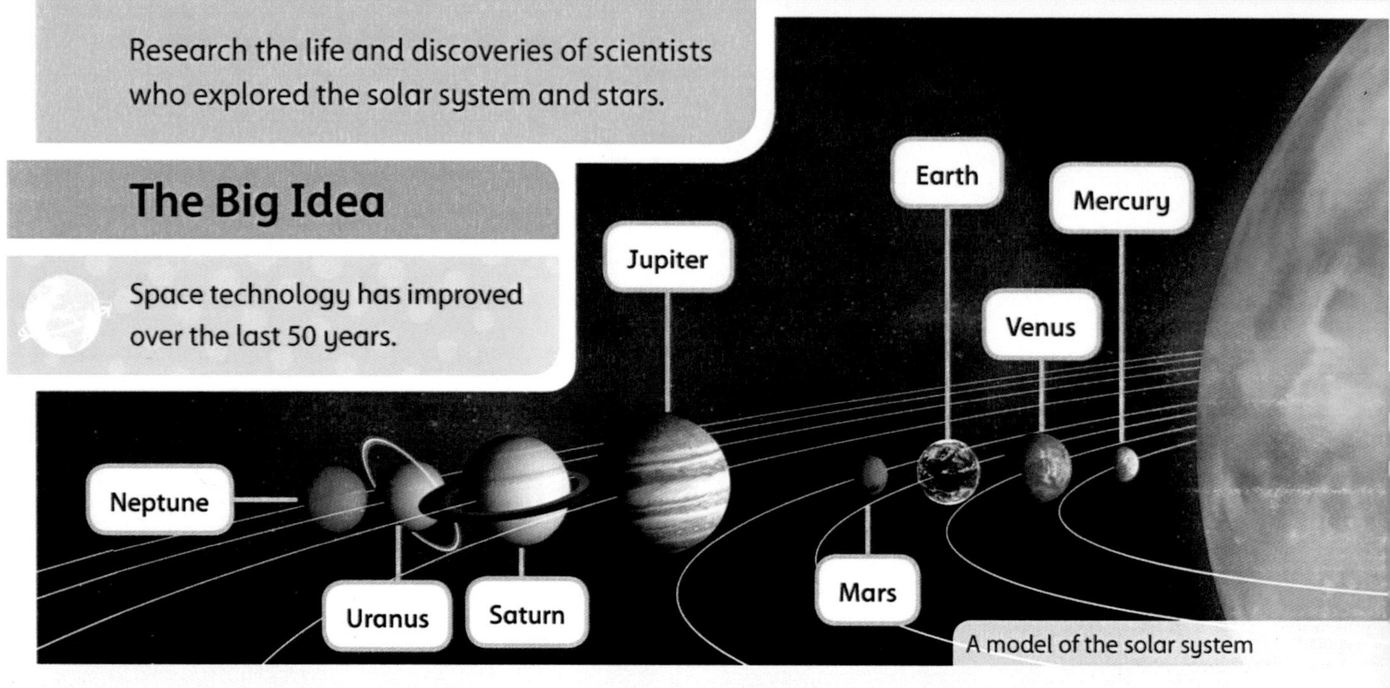

A model of the solar system

Hundreds of years ago scientists thought that the Earth was at the centre of the solar system. They could not explain why the Sun seemed to move across the sky. Then some scientists thought that the Sun was at the centre of the solar system.

Four hundred years ago an astronomer named Galileo built a telescope. He could then see the Moon, planets and stars more clearly. Galileo proved that the Sun was at the centre of the solar system.

 List the four planets closest to the Sun.

 Investigation: Modelling our solar system

You are going to make a scaled model of the planets of our solar system using a line constructed in your classroom to place the planets in the correct order.

1 Your teacher will give each group a planet, moon or star to create.
2 Using circular card, cut out a scaled model of your planet. Colour it so that it looks like the planet.
3 Hang your planet on the line in the correct place in relation to the Sun.

 List the eight planets in the correct order from the Sun, the closest first.

Amazing fact

Modern telescopes and **space probes** have shown us that the universe is made of billions of stars. These are grouped together in huge **galaxies**. Our Sun is in the Milky Way galaxy.

Think of your own way to remember the correct order of the planets.

It is important to know that when we look at planets in the night sky they do not make their own light. They are reflecting light from the Sun.

So far, there is no sign of life on any planet other than the Earth.

Planet	Distance from the Sun (million km)	Mean surface temperature (°C)
Mercury	58	170
Venus	108	460
Earth	150	15
Mars	228	-50
Jupiter	778	−143
Saturn	1427	−195
Uranus	2870	−201
Neptune	4497	−220

Look at the table above. Why might life not exist on any of the other planets?

Think about...

There are spaceships travelling to the outer reaches of the universe. They are unmanned because it takes so long to get there.

Telescopes are now much more efficient than Galileo's telescope. The Hubble telescope can see a distance of several billion light years. Space travel has also provided evidence that the Earth is spherical. There are hundreds of images and video clips of the Earth from space.

1 Is there any evidence to prove that the Earth is spherical?

2 Is the evidence reliable?

Now turn to page 119 to review and reflect on what you have learned.

What we have learned about Earth's movements

The Sun appears to move, but it doesn't (pages 102–9)

 What is the name given to how the Earth revolves or turns?

The Earth moves in two ways at the same time. It turns on its axis while moving around the …

I can model the movement of the Sun, Moon and Earth.

I understand how the Earth moves around the Sun.

How long does it take the Earth to spin on its axis? (pages 110–11)

Why does it get dark at night?

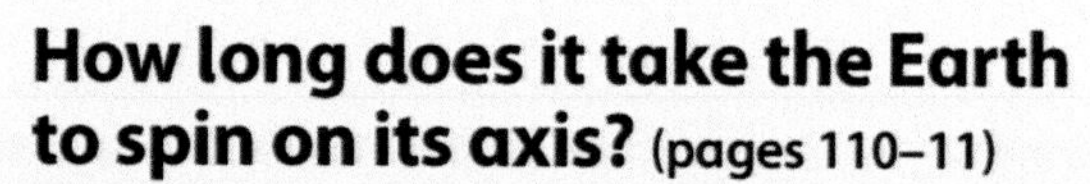

 Why do we cast hardly any shadow in the middle of the day?

I know how the Earth spins on its axis.

I know that it takes 24 hours for the Earth to make one complete spin.

I can make observations about the changes that take place over a day.

How does the Earth orbit the Sun? (pages 112–13)

How long does it take for the Earth to orbit the Sun?

 The Earth is tilted on its axis and for some of the year the northern hemisphere will be tilted away from the Sun. How does this affect the seasons?

I know it takes the Earth a year to orbit the Sun.

I know that the Earth spins and orbits at the same time.

I understand how we get summer and winter.

The discovery of our solar system (pages 114–17)

(pages 114–17)

Mercury, **Venus**, Earth, **Mars** and Jupiter are all planets in our solar system. Name the other three planets. (Hint: they spell 'SUN'!)

What is the name of the astronomer who proved through the use of a telescope that the Sun, not the Earth, was the centre of the solar system?

What shape is the Earth? Give two pieces of evidence to support this.

I know why the Earth is the only planet in the solar system to support life.

I understand that the Sun is at the centre of the solar system and the planets are arranged in inner and outer orbits.

I know what shape the Earth is.

I can explain how scientists can prove that the Earth is not flat.

6 Shadows

In this module you will:

- build on ideas about how light forms shadows
- investigate how shadows from the Sun change over time
- discover how light is measured.

Word Cloud

blocked
light intensity
length
position
silhouette

Before televisions and cameras were invented, people used to pay lots of money to have portraits painted. In the 1700s, Etienne de Silhouette invented a new way of making pictures of people. He invented the **silhouette**. These were much cheaper to produce than a painting.

How is the artist using a shadow to help him make a picture?

Why does the room have to be darkened?

Why is it better to use a lamp than a candle?

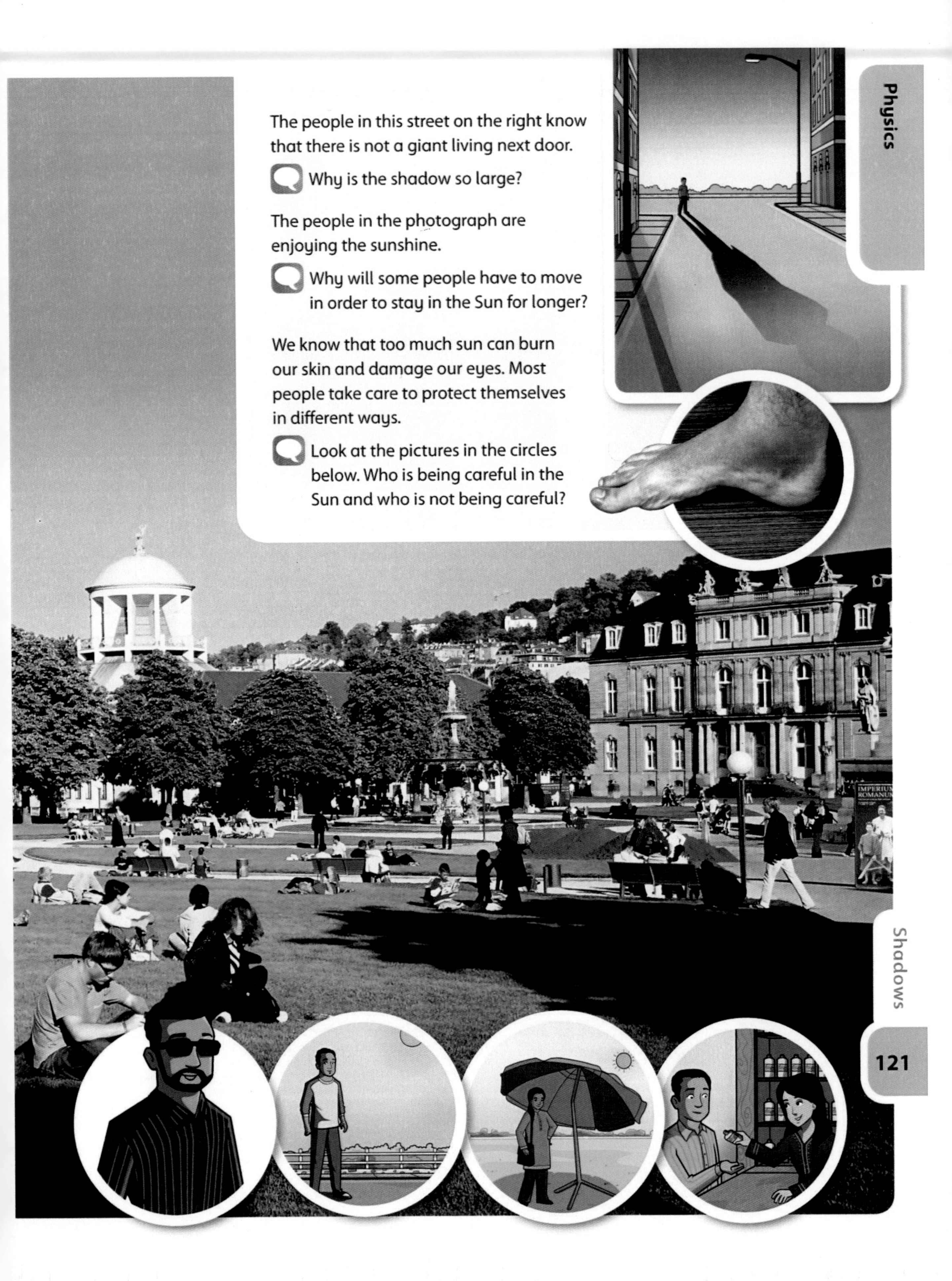

The people in this street on the right know that there is not a giant living next door.

Why is the shadow so large?

The people in the photograph are enjoying the sunshine.

Why will some people have to move in order to stay in the Sun for longer?

We know that too much sun can burn our skin and damage our eyes. Most people take care to protect themselves in different ways.

Look at the pictures in the circles below. Who is being careful in the Sun and who is not being careful?

Can we see through it?

Explore how transparent materials let a lot of light through and opaque materials do not let light through.

The Big Idea

Some substances let light through and other substances block the light.

 Write down two properties of light.

An object that does not let light through is opaque. Many objects are made of opaque materials. Your clothes and large parts of cars, buses and trains are opaque.

Why is it important that some materials are opaque? Imagine if we could see through every substance on Earth.

An object that lets a lot of light through is transparent. Many objects are made from transparent materials. We need to look through windows and see through air.

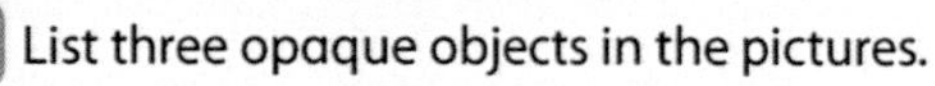

List three opaque objects in the pictures.

 List three transparent objects in the pictures.

Imagine all the materials where you live are opaque. What is life like?

Some materials let a little light through. These materials are translucent. We can see shapes on the other side of translucent materials but not very clearly. Coloured and frosted glass are examples of translucent materials.

Look at the pictures above. Is the transparent material labelled A, B or C?

Is the opaque material labelled A, B or C?

What can the person see through the translucent material?

Investigation: Materials

You are going to investigate some materials to find out if they are opaque, transparent or translucent.

1 Your teacher will provide you with some different materials to test and some objects to look at.

2 Before looking at the objects, predict whether each type of material is opaque, transparent or translucent.

3 Look at each object through the different materials. Decide whether each material is opaque, transparent or translucent.

4 Make a table in your Investigation Notebook to record your results. To make your results reliable, repeat your investigation. Think about this when drawing your table.

Are there any items that did not give you the results you predicted?

Are there any links between the uses of the materials and whether or not they let light through?

Do transparent materials block light?

Why is it important that not all of the materials we use are opaque?

Can we see through it?

Explore how transparent materials let a lot of light through and opaque materials do not let light through.

The Big Idea

We can link the property of a material with how good it is at making shadows.

These are shadows of windows. Which part of the shadow is the frame?

Which part of the shadow is the glass?

Which one of these materials is opaque?

A shadow forms when light is **blocked**. Opaque materials will not let light through. This suggests that opaque materials form the best shadows.

Does the photograph of the window shadows support this idea?

Investigation: Testing materials

You can investigate materials to test which make shadows. You need a light source and samples of different materials.

1 Set up your equipment.
2 Predict what you think will happen each time you try to cast a shadow using a transparent, translucent and opaque material.
3 Investigate which materials cast a shadow on the screen.
4 Record your results in your Investigation Notebook.
5 Compare your predictions with your results.

How accurate were your predictions?

Did opaque or transparent materials make the best shadows?

 Are the shadows made by the opaque and translucent materials the same?

Opaque materials can be very useful. We sit under sunshades or use umbrellas to keep the Sun off babies and young children.

Imagine you work for a sunshade company. Your job is to design a sunshade for the company to sell. You need to decide what type of material to use.

Investigation: Testing sunshades

1 In your group, plan how you will set up this investigation.
2 Discuss your plan with the rest of the class before you start the investigation.
3 Use a light to represent the Sun. To make sure that this is a fair test, fix the light in **position** so it does not move and stays at the same height.
4 Hold each material in turn below the light and observe the shadow it casts.
5 Record your results in your Investigation Notebook.
6 Compare your results with those from the rest of the class.

 1 Why is it important for some materials to be transparent?

2 Why it is important for some materials to be opaque?

3 List two uses of opaque materials.

4 How can you make a shadow larger?

Now turn to page 138 to review and reflect on what you have learned.

Creating shadows

Observe that shadows are formed when light is blocked by opaque materials.

The Big Idea

Shadows can be used for fun.

Investigation: Playing a shadow game

1. Your teacher will give everyone a torch and the room will be darkened.
2. Select one person in the class to start the game. They choose an object in the room and begin to describe it without looking at it.
3. Everyone else uses their torch to find the object. The first person to find the object is the next one to play.

Investigation: Making shadow puppets

You have learned a lot about shadows and light. Now you are going to write and perform your own short shadow puppet show.

1. Working in small groups, plan and write a short shadow puppet show to perform to the rest of the class.
2. In your group, decide who the main characters are in your puppet show.
3. Using the materials provided, make puppets for your main characters. Think back to all the investigations you have carried out. This will help you to decide what size the puppets should be.
4. Practise positioning the puppets in front of the light source.
5. In your group, perform your puppet show to the rest of the class.

How can you make your puppet show more colourful?

You could also perform behind a sheet of fabric or a blind to make the images look different.

Investigation: Making 3D glasses

You can make 3D glasses using a strip of card and blue and red cellophane. These make flat pictures look more real.

1 Mark the card where each eye will be.

2 Cut out a 3 centimetre by 2 centimetre square for each eye.

3 Cover one eye hole with blue cellophane and the other with red. Stick the cellophane in place.

4 When you look at special pictures with your 3D glasses the images will appear to be 3D.

Which two colours should you use for the lenses?

Investigation: Student silhouettes

1 Shine a light at the side of your face. Ask a friend to tape some paper to the wall where your shadow appears.

2 Ask them to draw around your shadow on the paper and cut it out. You now have a silhouette of your head.

3 Make this into a shadow puppet and you can be in your play.

1 Why do shadow puppets have to be made from opaque materials?

2 Which two colours are needed to form a 3D image?

3 Describe how to make a silhouette.

Now turn to page 138 to review and reflect on what you have learned.

Growing and shrinking shadows

Investigate how the size of a shadow is affected by the position of the object.

The Big Idea

Shadows can grow bigger and smaller.

Think back to your shadow puppets. What happened to the shadows when you moved the puppets closer to the screen?

Investigation: The size of a shadow

What do you think will happen to the shadow if we move an object closer to the light source?

1 In groups, carry out an investigation to explore if your prediction is true.

2 Using a torch as your light source, place it 2 metres away from a screen or wall and fix it in place so it does not move.

3 Choose an opaque object that is easy to move. You will need to measure this object, so choose something that has a simple shape, such as a building block.

4 Place the object 10 centimetres in front of the light source and measure the **length** of its shadow.

5 Now measure the size of the shadow from the same object at 20 centimetres, 30 centimetres, 40 centimetres and 50 centimetres away from the light source.

Will these measurements be reliable?

What should you do next to make sure the results are reliable?

6 In your Investigation Notebook, record your results in a table.

Which table below would be the better one to use? Think! Which will allow you to collect repeat results?

Distance from light source (cm)	Length of shadow (cm)

Table A

Distance from light source (cm)	Length of shadow (cm) Try 1	Length of shadow (cm) Try 2	Length of shadow (cm) Try 3

Table B

What do you notice about your results? Can you see a pattern?

Was your prediction at the start of the investigation correct?

Which variables were kept the same in the investigation to make sure it was a fair test?

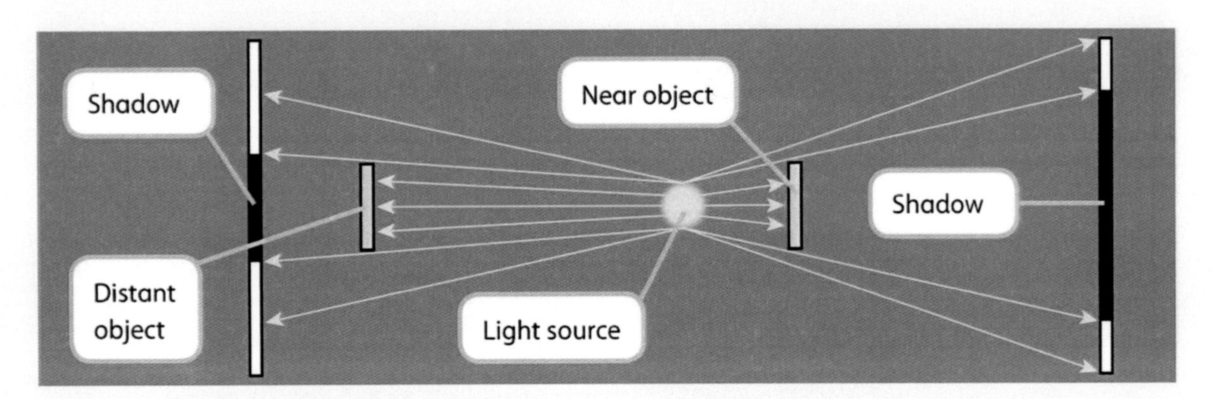

Look at the diagram above. When the light source moves closer to an opaque object the shadow gets bigger. If we move the light source away from an opaque object, the shadow gets smaller. This is because light beams travel in straight lines. They cannot pass through the opaque material and they cannot bend around it.

How can you make your shadow puppet look like it is moving without actually having to touch it?

How can you make your shadow puppet look larger or smaller?

1 Circle the type of material that must be used to make a shadow puppet:

Transparent   Opaque   Translucent

2 Why does the shadow of an object get bigger as the object is moved closer to the light?

3 How can you collect reliable data?

Now turn to page 139 to review and reflect on what you have learned.

Tracking those moving shadows

Observe that shadows change in length and position throughout the day.

The Big Idea

Shadows change throughout the day.

What is the best natural source of light?

List two artificial sources of light that you have used in investigations so far.

During the day, the Sun appears to move across the sky. In the morning, the Sun is lower on the horizon and is not as powerful. At midday, the Sun is much higher in the sky and is at its most powerful. This makes the light seem brighter and it is hotter. After midday, the Sun gets lower towards the horizon again and is much less powerful and cooler than at midday.

Investigation: The apparent movement of the Sun

1 Take your shadow puppet outside to make shadows or simply use yourself.

2 Change position during your investigation. Move your arms or legs. Try running on the spot or hopping.

3 Observe how your shadow moves. Does it move exactly like you? Draw around your shadow with chalk or mark it using pegs to show how it changes throughout the day. Record the time.

4 Keep a diary of your shadow. Check how long it is in the morning, at midday and in the afternoon.

5 In your Investigation Notebook, record your results in a table like the one below.

6 Observe how your shadow changes day to day and month to month.

Date	Time of day	Observations and length of shadow

It might not be possible to finish this investigation in one day, but try to record your shadow every hour or as often as you can. Also, look at the shape of your shadow.

Does your shadow change shape?

Does your shadow change in size?

Does the darkness of your shadow stay the same?

Write a conclusion about your investigation.

During the morning my shadow is …

At midday my shadow is …

During the afternoon my shadow is …

Think about...

When planning buildings, architects have to think about how the Sun will appear to move and how this will alter the light and shade. The building may be in the shadow of another building at one part of the day, but in bright sunlight later in the day. How do moving shadows impact on the light and shade where you live?

Tracking those moving shadows

Observe that shadows change in length and position throughout the day.

The Big Idea

We can use the size and direction of a shadow to tell the time.

Look at the object in the picture below. What is it? What is it used for?

The shadow cast by the Sun will change throughout the day.

Investigation: Shadow sticks

1 Find a place outside and push an upright stick into the ground.

2 Draw around the shadow of the stick at the start of your investigation. You can do this on paper or use chalk on the ground.

3 Go back to your stick every hour and mark the new position of the shadow so that you can measure its length.

4 Record your results in a table in your Investigation Notebook.

What do you notice about the length of the shadow?

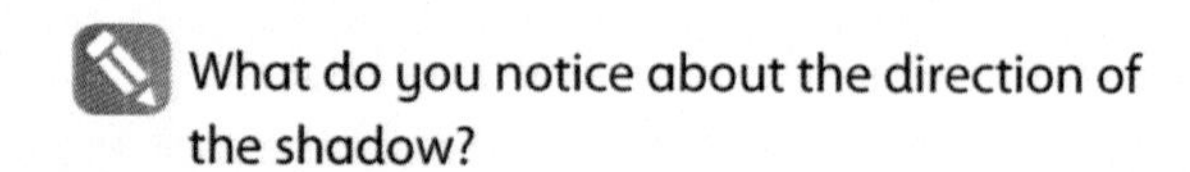

What do you notice about the direction of the shadow?

You could repeat this experiment for a day every month.

 What do you predict will happen to the shadows?

It is possible to find out the exact time of midday with the results from this investigation. When the shadow is at its shortest, the Sun is directly above you. This is midday.

 Investigation: Tracking your shadow

1 Work in groups of three or four and find a clean, smooth and safe surface outside.
2 Make a mark on the ground so that you stand in the same position every time. You could draw around your feet.
3 Stay in one position and stand very still. The other people in your group use chalk to draw around your shadow on the ground. (You may prefer to use paper.) Make sure nobody erases the diagram of your shadow.
4 If possible, revisit the drawing of your shadow every hour and draw around your new shadow. Over the course of the day, your shadow should change in size and position.

 On the picture above, label the shadow that was drawn at midday. Explain your decision.

Try to check if this changes over a week or even a month.

What do you predict will happen to the length of your shadow over a month?

 Decide whether the following statements are true or false.

		True/False
1	The Sun is the only natural source of light.	________
2	My shadow did not change at all throughout the day.	________
3	My shadow was at its longest at midday.	________

Now turn to page 139 to review and reflect on what you have learned.

Measuring light intensity

Know that light intensity can be measured.

The Big Idea

Light sources vary in brightness and it is vital that we can measure this accurately.

We can describe how bright or intense a light source is by using words such as very bright, dim, dark or glowing. It might even be possible to compare different light sources. We know that the Sun is brighter than a candle and a lighthouse is brighter than a torch. However, this still isn't very scientific. Scientists need to measure **light intensity** much more accurately.

Light intensity is how much light energy there is in one place. Think about the ray box you made in Module 1, *The Way We See Things*.

Directing the light through the tiny slit makes the light seem brighter or more intense.
The wider the slit, the less intense the light.
Brightness and light intensity are similar, but what appears bright to one person may not seem as bright to another person. We need a way of measuring light intensity.

A photodiode

A photodiode changes the light energy into an electrical current. This is displayed as a number. These devices are sometimes called light meters.

 Why is it better to have a number to measure light intensity rather than using words?

Light meters are used to measure the strength or brightness of electric light bulbs. Electricians also use them to measure the light intensity in a room. Some rooms need to have more light intensity than others.

Hospital operating theatres need a very high light intensity. A bedroom does not need a high light intensity.

 Why does an operating theatre have to have a high light intensity?

Light meters are also used in digital cameras and mobile phone cameras. They measure the light intensity around an object. If there is not a high enough reading, the flash is used.

How do our eyes change the amount of light that enters?

Why can't we see very well in a low light intensity?

Have you ever been on a road or in a big building and the lights have come on automatically? This is because a light meter has recognised that there isn't enough light.

Look carefully at a solar calculator and notice the strip of solar panel. This transfers light energy to electrical energy to enable us to use the calculator.

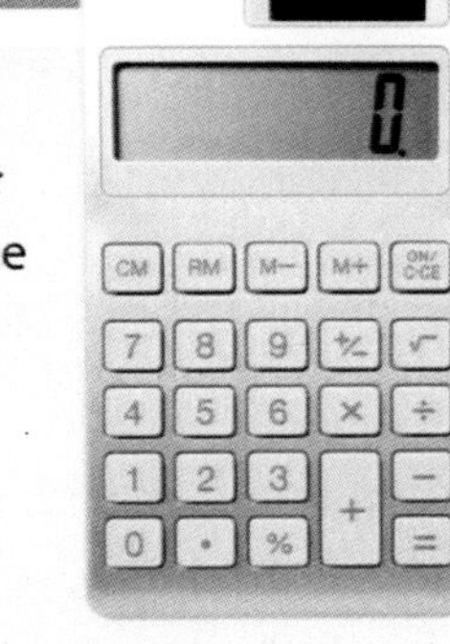

What happens if you cover the strip?

Investigation: Measuring light intensity

We can measure how much light a calculator needs to work.

1 Place the calculator beneath a bright light source. A small table lamp or torch fixed in place will work well.
2 Switch the calculator on and enter the number 1000 to check that it is working.
3 Cover the solar panel with one piece of tissue paper. Check to see that the number is still showing.
4 Record this in a table of results in your Investigation Notebook.
5 Keep adding layers of tissue paper one at a time and filling in your results table.
6 Record the number of layers of tissue paper needed to stop the calculator from working.

How can the calculator and tissue paper be used to measure the light intensity of different light sources?

Plan an investigation to compare a torch, a candle, the Sun and a table lamp.

Measuring light intensity

Know that light intensity can be measured.

The Big Idea

Scientists have been curious about light and have carried out experiments to help us to understand it.

Why do we need to measure light intensity?

What might happen if we have no way of measuring the brightness of light sources?

Measuring the light intensity of a car headlight

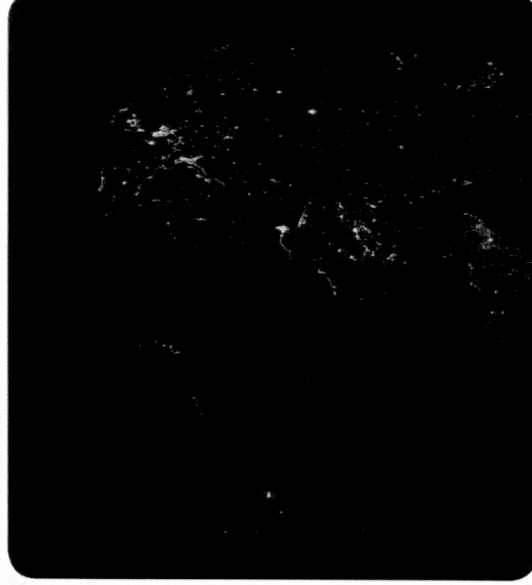

Before the 18th century, not many people measured the intensity of light. Some astronomers tried to use light levels to estimate how far away planets were from the Earth.

At this time, light intensity was measured in foot-candles. This is a very old English method of measuring light intensity.

During the 19th century, the electric light was invented. People tried to prove that the electric light was more efficient than the gas light. In 1920, light intensity was measured in an accurate way by using luminance. Luminance is the total amount of visible light present and is measured in lumen.

Light intensity is also measured in a unit called lux. 1 lux is equal to 1 lumen.

When building offices and factories, there are strict rules about light intensity. If the light is too bright, workers might suffer from headaches and the cost of electricity will be high. If the buildings are too dark, they can be dangerous and workers may not be able to see to work properly.

There are fewer rules when it comes to our own homes. It is still important to make sure there is enough light, though, to see clearly.

Activity	Foot-candles	Lux
Washing clothes	10–15	100–150
Eating	20–30	200–300
Reading/ studying	50–150	500–1500
Sewing	100–200	1000–2000

Light intensity levels linked to activities at home

Look at the results in the table above. Which two activities need the most light? Why?

If you have ever stood in a darkened room with just one source of light, have you noticed how the light fades as you move away from the light?

The table below shows some results from an investigation to prove this.

Distance from source (cm)	Light intensity (lux)
10	930
15	410
20	235
25	155
30	100
50	40
70	20
80	15
100	10

Distance and light intensity

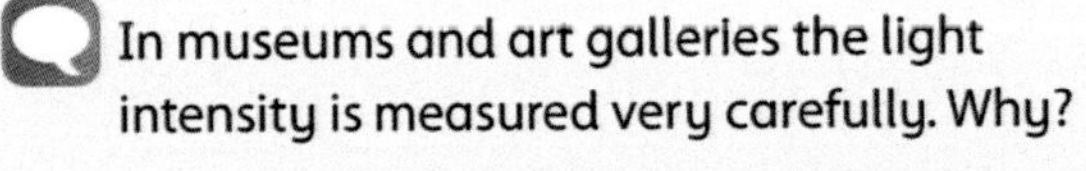

In museums and art galleries the light intensity is measured very carefully. Why?

Draw a chart in your Investigation Notebook to show these results.

1 Why do scientists measure light intensity?

2 What equipment is used to measure light intensity?

Now turn to page 139 to review and reflect on what you have learned.

What we have learned about shadows

Can we see through it?
(pages 122–5)

What word do we use to describe substances that let lots of light through? Give three examples.

I know that transparent materials let a lot of light pass through them.

I know that translucent materials let some light pass through them.

What word do we use to describe substances that let some light through? Give three examples.

What word do we use to describe substances that do not let light through? Give three examples.

I know that opaque materials do not let light pass through them.

Creating shadows (pages 126–7)

How can you make a puppet shadow that is very dark?

How can you make a coloured shadow?

I know that shadow puppets must be made from opaque materials.

I can use shadows to make a puppet play.

Growing and shrinking shadows (pages 128–9)

How can you make a puppet shadow grow larger on the screen?

Why do scientists repeat investigations to get another set of results?

I know why an object's shadow gets bigger as it is moved nearer to a light source.

I can plan to collect reliable results.

Tracking those moving shadows (pages 130–3)

What time of day is your shadow the shortest?

What time of day is your shadow at its longest?

When the shadow changes its length and position, does the Sun move or does the Earth move?

I can explain why a shadow changes position throughout the day.

Measuring light intensity (pages 134–7)

Why do scientists need to measure light intensity?

What is the modern unit of light measurement called?

I know two places where a light meter is used.

I know that all light travels at the same speed and that brightness depends on light intensity.

Glossary

Key words

24 hours

axis

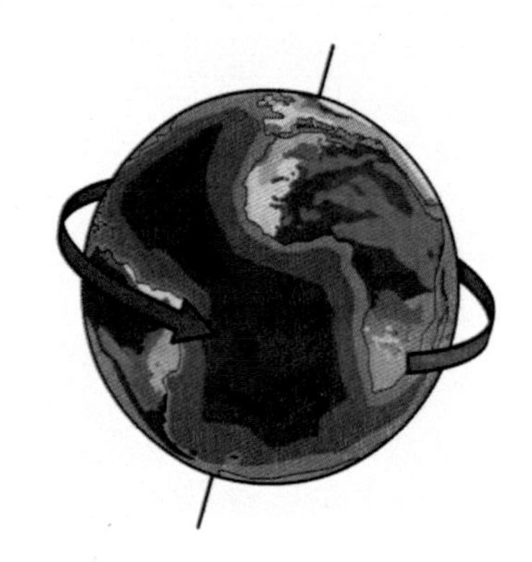

beam

blocked

boiling point

condensation

crystal

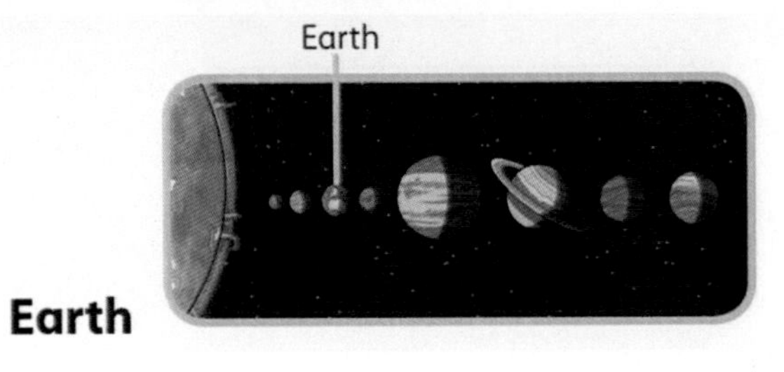

Earth

depict

energy

dispersal

evaluate

draw conclusions

evaporation

explosion

female (ovum)

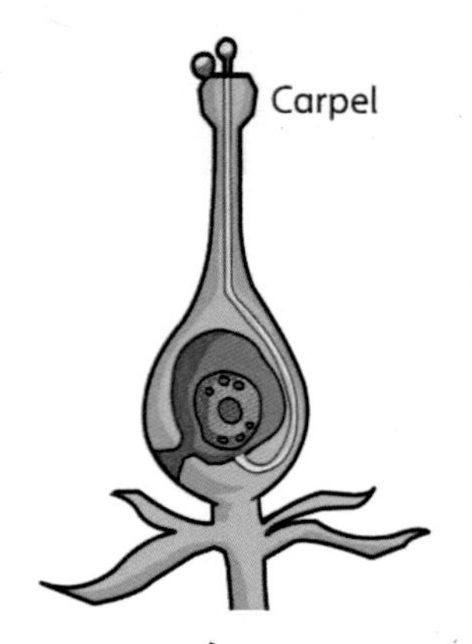

fertilisation

filter funnel

filter paper

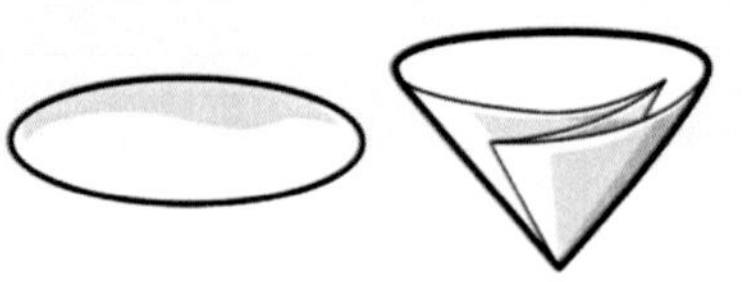

fruit

galaxy

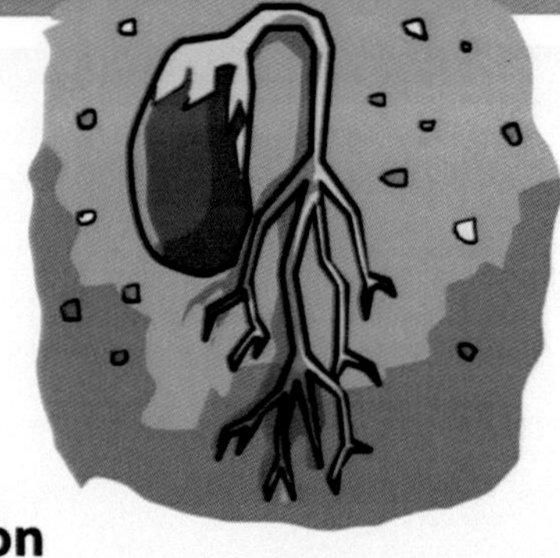

germination

insect

insoluble

interpret

interpret data

interpret findings

Jupiter

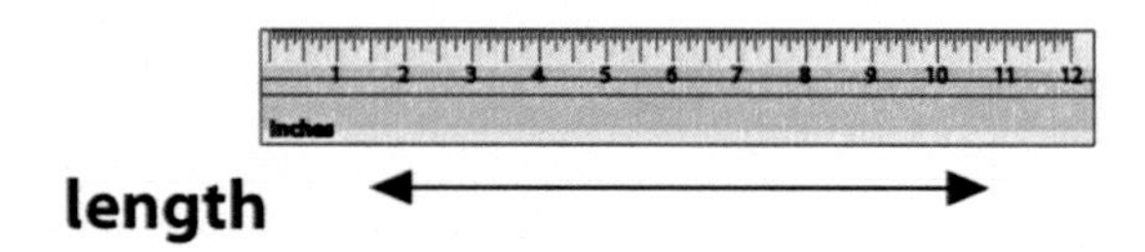

length

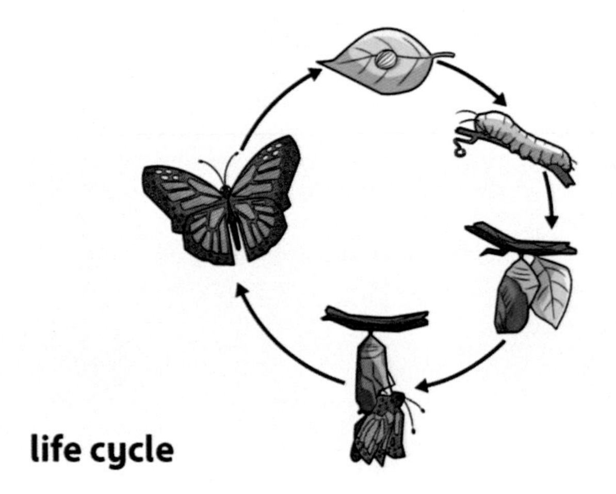

life cycle

light energy

light intensity

light source

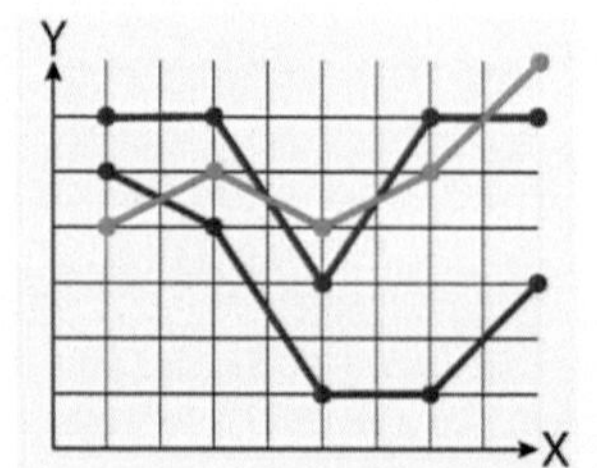

line graph

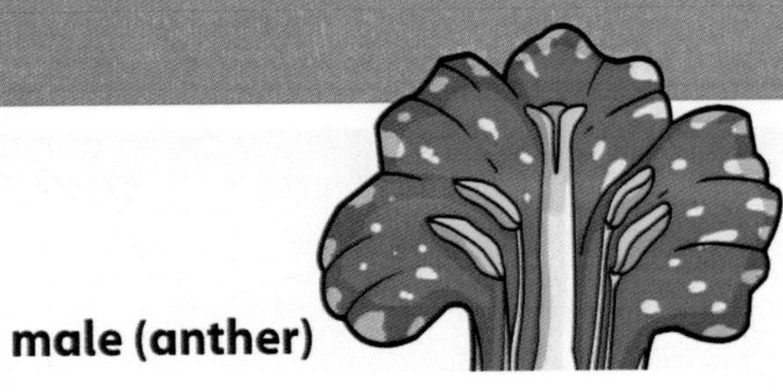

male (anther)

Mercury

Mercury

Mars

mirror

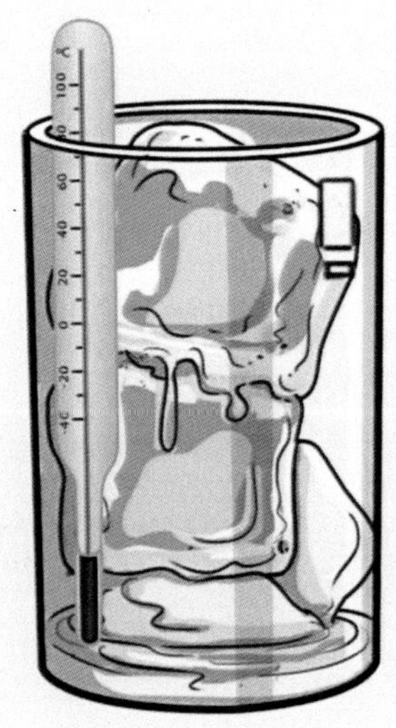

melting point

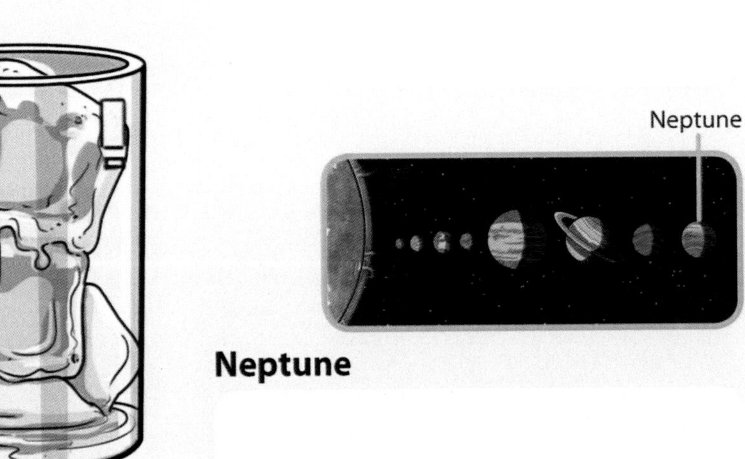

Neptune

object

opaque

pollination

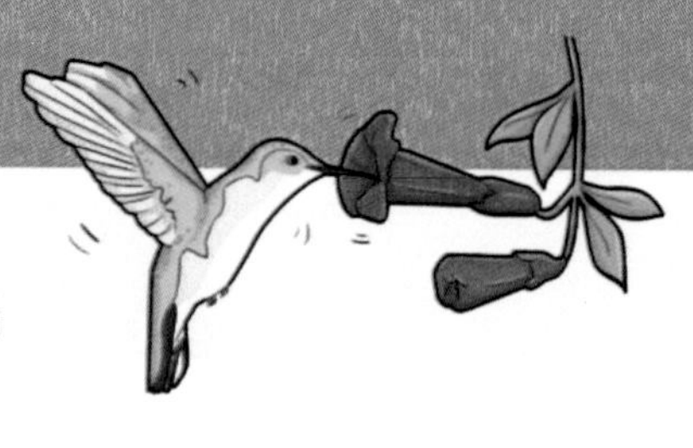

position

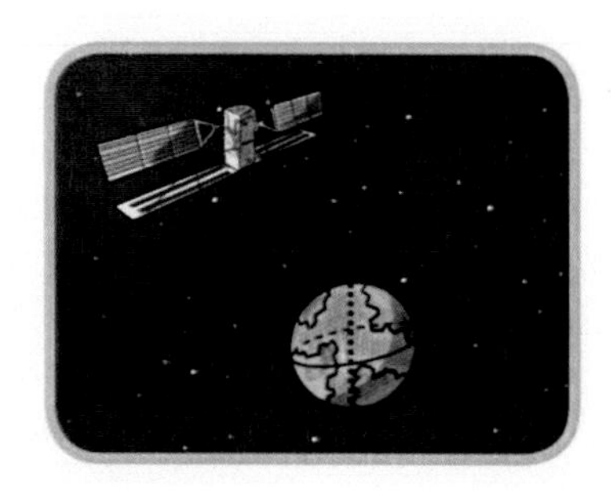

orbit

present

planet

ray

recognise patterns in data

Saturn

reflect

repeat measurements

research

scientist

seed production

sieve

silhouette

soluble

state of matter

solution

surface

translucent

space probes

transparent

star

Venus

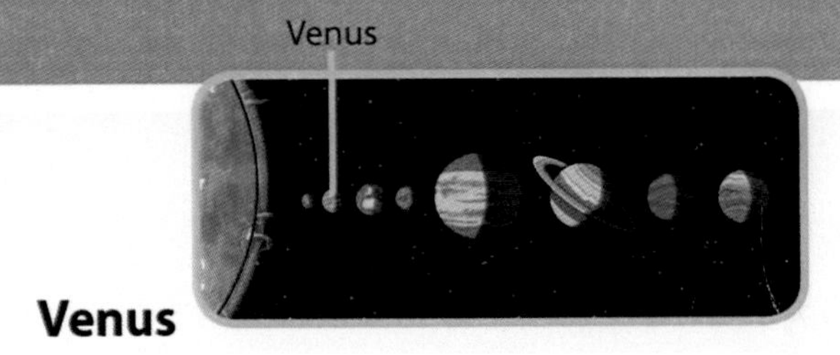

water vapour

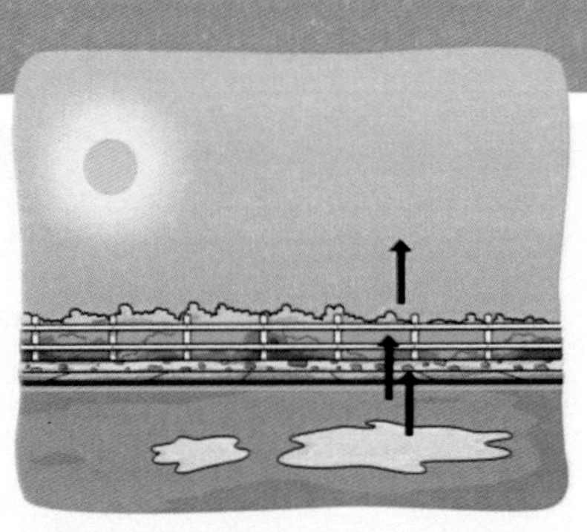

warmth

wind

water cycle

year

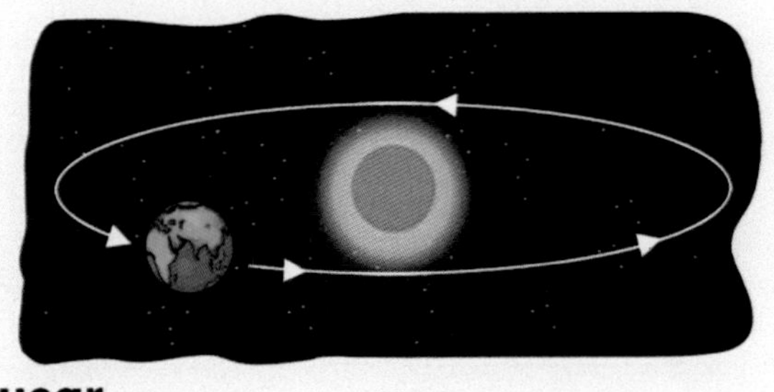